Emily Dickinson's Poetic Art

COGNITION, POETICS, AND THE ARTS

The **Cognition, Poetics, and the Arts** series fosters high-quality interdisciplinary research at the intersection of the cognitive sciences and the arts that focuses on cognitive approaches to literatures, arts, and cultures from around the world with three major objectives: (1) to develop theories and methodologies that further our understanding of the arts as central and complex operations of human minding; (2) to investigate the ways models of minding and artistic creation and reception have been developed and revised in relation to each other throughout history and in different cultural contexts; and (3) to develop theoretical and methodological understandings of how the arts illuminate and contribute to the cognitive sciences.

Series Editors

Alexander Bergs, University of Osnabrück, Germany
Margaret H. Freeman, Myrifield Institute for Cognition and the Arts, USA
Peter Schneck, University of Osnabrück, Germany

Volumes in the Series:

Emily Dickinson's Poetic Art: A Cognitive Reading, by Margaret H. Freeman
Rhythm in Modernist Poetry: An Essay in Cognitive Versification Studies, by Eva Lilja
(forthcoming)
Cognition in the Poem: Processes of Subjectivity in Comparative Poetics, by Victor
Bermúdez (forthcoming)

Emily Dickinson's Poetic Art

A Cognitive Reading

Margaret H. Freeman

BLOOMSBURY ACADEMIC
NEW YORK · LONDON · OXFORD · NEW DELHI · SYDNEY

BLOOMSBURY ACADEMIC
Bloomsbury Publishing Inc
1385 Broadway, New York, NY 10018, USA
50 Bedford Square, London, WC1B 3DP, UK
29 Earlsfort Terrace, Dublin 2, Ireland

BLOOMSBURY, BLOOMSBURY ACADEMIC and the Diana logo are trademarks of
Bloomsbury Publishing Plc

First published in the United States of America 2023

For legal purposes the Acknowledgments on p. xi constitute an extension
of this copyright page.

Series design: Tjasa Krivec
Cover image: "Solarcan Matrix, Week 5." Williestruther Loch captured with a Solarcan.
© Sam Cornwell, 2019

Bloomsbury Publishing Inc does not have any control over, or responsibility for, any
third-party websites referred to or in this book. All internet addresses given in this
book were correct at the time of going to press. The author and publisher regret any
inconvenience caused if addresses have changed or sites have ceased to exist,
but can accept no responsibility for any such changes.

Library of Congress Cataloging-in-Publication Data
Names: Freeman, Margaret H., author.
Title: Emily Dickinson's poetic art : a cognitive reading / Margaret H. Freeman.
Description: New York : Bloomsbury Academic, 2023. | Series: Cognition, poetics, and
the arts | Includes bibliographical references and index. | Summary: "An exploration of
both a major American poet, Emily Dickinson, as well as cognitive approaches to literary
criticism"– Provided by publisher.
Identifiers: LCCN 2022046415 (print) | LCCN 2022046416 (ebook) | ISBN 9781501398193
(hardback) | ISBN 9781501398186 (paperback) | ISBN 9781501398209 (eBook) |
ISBN 9781501398216 (ePDF) | ISBN 9781501398179 (electronic)
Subjects: LCSH: Dickinson, Emily, 1830-1886–Criticism and interpretation. | Cognition in
literature. | LCGFT: Literary criticism.
Classification: LCC PS1541.Z5 F745 2023 (print) | LCC PS1541.Z5 (ebook) |
DDC 811/.4–dc23/eng/20230210
LC record available at https://lccn.loc.gov/2022046415
LC ebook record available at https://lccn.loc.gov/2022046416

ISBN: HB: 978-1-5013-9819-3
PB: 978-1-5013-9818-6
ePDF: 978-1-5013-9821-6
eBook: 978-1-5013-9820-9

Series: Cognition, Poetics, and the Arts

Typeset by Deanta Global Publishing Services, Chennai, India
Printed and bound in the United States of America

To find out more about our authors and books visit www.bloomsbury.com and
sign up for our newsletters.

For All
Dickinsonians
Everywhere

Contents

Figures

Preface

The availability of original manuscripts online has revolutionized the way we may encounter an Emily Dickinson poem. Since Dickinson did not oversee her poetry into print publication, editors from the very beginning have been faced with the enormous and problematic task of rendering her written manuscripts into manageable and readable form. As readers, we owe them an enormous debt in making Dickinson's poetry accessible to the world at large. The greatness of her reputation is due in large part to their efforts.

It is not my intention, therefore, to belittle their efforts or the many scholarly works that have taught us so much about the life and works of a great poet. Rather, in focusing on the manuscripts themselves, I introduce a new way of reading, one that employs the theories of the cognitive sciences and aesthetics to enable us to cognitively experience Dickinson's poetry. The cognitive revolution that took place in mid-twentieth century upended many presuppositions that arose out of mainstream Western tradition: namely, (1) the Cartesian body-mind split that led to an objectivist view of our place in the world; and (2) the notion that cognition was a purely mental operation involving rational and logical reasoning as opposed to the emotive, sensory, and motor functions of the body-self. Cognition is now understood by many researchers in both the sciences and the arts as being embodied, embedded, enactive, and extended.

The consequence of this cognitive shift results in reading poetry, not from the viewpoint of interpretation only, but through experiencing the physical and subliminal affective manifestations of a poet's and readers' own cognitive processes. That is what this book aims to do in giving fresh readings of Dickinson's poetry.

Acknowledgments

I owe an enormous debt of gratitude to so many over so many years during my long journey of discovery in better understanding the wellsprings of Emily Dickinson's poetry. Especial thanks go to Mike Kelly, the Amherst College archivist, for his help in making the manuscripts accessible and for sending me high-resolution images; to Leslie Morris and Christine Jacobson at the Houghton Library for their willingness to help on all manuscript matters; to Emily Seelbinder for her Dickinson expertise in reading carefully each chapter; to Pat McGahan, whose computer expertise reformatted and rendered my figures in correct resolution; to Lynn Perry and Anne Williamson for commenting on the readability of several chapters; and to my most patient spouse Donald for his valued commentary on the earliest draft.

Credits

Emily Dickinson Collection, Amherst College (A)
  https://acdc.amherst.edu/collection/ed

Houghton Library, Harvard University (H)
  https://library.harvard.edu/collections/emily-dickinson-collection

All quotations of the poems are transcribed from the manuscript versions. Poems are identified by first line and numbering from the three major editions: R. W. Franklin (F), T. H. Johnson (J), and C. Miller (M). Quotations from the Johnson edition of letters are marked with an L.

David Hinton's epigraph in Chapter 9 is given with kind permission of the author.

Material reproduced by permission with revision from previous publications are as follows.

Chapter 2: "The Body in the Word: A Cognitive Approach to the Shape of a Poetic Text." In *Cognitive Stylistics: Language and Cognition in Text Analysis*, edited by Elena Semino and Jonathan Culpeper, 23–47. John Benjamins, 2002. https://benjamins.com/catalog/lal.1
https://benjamins.com/catalog/lal.1.04fre

Chapter 8: "Emily Dickinson and the Discourse of Intimacy." In *Semantics of Silences in Linguistics and Literature*, edited by Gudrun M. Grabher and Ulrike Jeßner, 191–210. Heidelberg: Universitätsverlag C. Winter, 1996.

Chapter 9:   "Grounded Spaces: Deictic -Self Anaphors in the Poetry of Emily Dickinson." *Language and Literature* (1997), 6 (1): 7–28.

Chapter 11:   "Cognitive Mapping in Literary Analysis." *Style* (2002), 36 (3): 466–83. Used with permission from Penn State University Press.

Chapter 12:   "Reading Readers Reading a Poem: From Conceptual to Cognitive Integration." *Cognitive Semiotics*, Issue 2 (Spring 2008): 102–28.

Chapter 13:   "Metaphor Making Meaning: Dickinson's Conceptual Universe." *Journal of Pragmatics* 24 (1995): 643–66. ISSN 0378-2166, https://doi.org/10.1016/0378-2166(95)00006-E. (https://www.sciencedirect.com/science/article/pii/037821669500006E)

Chapter 14:   "Poetry and the Scope of Metaphor: Toward a Cognitive Theory of Literature." In *Metaphor and Metonymy at the Crossroads: A Cognitive Perspective*, edited by Antonio Barcelona, 253–81. Berlin and New York: Mouton de Gruyter, 2000.

Demure as Dynamite

Dickinson and Cognition

It has been said that the mystery of Mozart "is the very nature of his ordinariness"; that if you "take the music away . . . what's left? Not an exceptional life."[1] One can hardly imagine a similar statement being made about Emily Dickinson. Her life intrigues as much as her poetry. Since Lavinia first brought those voluminous packets of poems to light after her sister's death in 1886, people have puzzled over the extraordinary quality of the poet's life and work. This fascination with Dickinson continues into the present. The most remarkable aspect of her reception over the years has been the enormous number of questions raised about both her life and poetry and the large and diverse range of "answers" proposed. As we attempt to understand the power and endurance of Dickinson's poetry, we find ourselves still raising more questions than answers.

More than any other poet, perhaps, Dickinson inspires a singular lack of unanimity among critics with respect either to what she was like as a person or to the meanings and qualities of her poems. Was she religious? A mystic? A skeptic? As a poet, where to place her? A Romantic? Victorian? Modern? A writer of confessional poetry? Critics have argued for all these and more.[2] Her art raises no fewer questions than her life. When were the poems written? Can meaning be attached to her grouping of poems in packets (called "fascicles" by Mabel Loomis Todd)? What does one do with the variants and alternate copies of a particular poem? Or her seemingly erratic capitalization and punctuation? Is she ungrammatical or simply complex? What do her poems mean or signify? Why, we finally ask in some exasperation, does Dickinson lend herself so readily to the possibility of finding in her poems whatever it is we are looking for?

To all these questions, we attempt to find answers. But in doing so, we sometimes trap ourselves into believing that somehow "the" answer will bring us to the goal of complete understanding. We turn ourselves into detectives of classical whodunits,

---

[1] The remarks were made by Peter Shaffer, author of *Amadeus*, and its director Sir Peter Hall in an interview with Nan Robertson (1980: C25).

[2] These questions were already being raised by the time of the one hundredth anniversary of Dickinson's death. Reviews of Emily Dickinson biography and critical commentary reveal the range of associations made in attempting to place the poet within a tradition (Buckingham 1970; Duchac 1979). In the many years since, interest in Dickinson's inscrutability has grown exponentially, not only among literary scholars but through her representation in other arts and social media.

concentrating on the solution of her mystery at the expense of a somewhat different focus: on the fact itself of the multitudinous questions Dickinson raises, and the evasions, elisions, and ambiguities: the enigma of Dickinson's life and art. So if Mozart's mystery is his ordinariness, Dickinson's is her inscrutability. The very reclusiveness that led her to withdraw from social contact makes it difficult for us to see her clearly, even now, after so much more has been revealed about her life and art in the many volumes on the poet that have appeared in the last half century or so.

So the question becomes what this signifies for our understanding of Dickinson's stature as a poet. Although linguistic analysis can reveal certain patterns of her semantic, syntactic, and prosodic practices, it cannot explain the inscrutable quality of her poetry. There is, first, the enigma of her variants, the seeming inability sometimes to settle on one word at the expense of another, the lack of ostensible referentiality of words and pronouns. There is her propensity for reversing the order of phrases, the seeming ungrammaticality of her sentences with their frequent ellipses. The elements of her prosody, the line breaks, capitalizations, punctuation markings, and metrical irregularities create effects that are not easy to explicate.

The question then is whether enigmatic qualities such as these detract from or enhance the poetry. An early critic, Thomas Bailey Aldrich (1892), called her poems *disjecta membra* because of their apparent ungrammaticality, and concluded as a result that "oblivion lingers in the immediate neighborhood."[3] It may be because of this very characteristic that the opposite is true. For despite the negativity evinced by some of her earliest critics, Dickinson's poetry has achieved both endurance and universality. By 1898, she had already been translated into German, followed in rapid succession by Dutch, Italian, Hungarian, Russian, Hebrew, and Japanese, to name just a few (Buckingham 1970, 165–201). She is read and loved the world over, by academics and non-academics alike, by people of different backgrounds and cultures, from housewives to Trappist monks.[4]

The questions are endless, but what emerges as a result of such multiple possibilities of interpretation is that each reader feels "ownership" of Dickinson, more directly, perhaps, than is the case for other poets, a position reflected in Susan Howe's (1985) book title, *My Emily Dickinson*. This feeling of ownership explains Dickinson's enormous popularity and endurance, which enables people of different ages and cultures to see her as somehow speaking for themselves.

So, in the end, it is more a question of experiencing, rather than interpreting, Dickinson's poetry. That is, in reading a poem, we have a tendency to filter its words through our own presuppositions and concerns which then constrain and determine the way we interpret their meaning. In doing so, we are "inter-fering"—carrying our own selves into the poem's "self" or "being," rather than letting the poem reveal

---

[3] Aldrich probably had in mind Horace's phrase *disjecti membra poetae*: "limbs of a dismembered poet."

[4] At the Emily Dickinson sesquicentennial celebration held at Amherst, Massachusetts (October 3–4, 1980), a member of the audience identified herself as a housewife who shared a love of Emily Dickinson's poetry with her Trappist brother who, with his brother monks, uses the poems for meditation.

itself. Culture also may constrain the way we read. In working with Masako Takeda, a Dickinson friend and colleague who translates Dickinson into Japanese, I was startled to learn that her original analysis of *A sloop of amber* (A112, F1599/J1622/M642) referred to the sunset in terms of a big and little cloud with no mention of the sun as ship (Freeman and Takeda 2006). It made me realize how distinct cultures may cognize literary experience quite differently.

Two approaches I've read recently, one philosophical, the other literary, have inspired my thinking along these lines. Gudrun M. Grabher (2019) introduced me to the works of Emmanuel Levinas, who argues that ethics is the first principle of philosophy: "It is the Other, not I, that is undefinable. The Other produces responsibility in me: the responsibility to learn from the Other and to answer not just for myself but also for the Other" (quoted in R. Gibbs 1991, 219). The second approach comes from my readings in Chinese classical poetry, which focuses on the need to empty one's mind from our own cultural assumptions, to experience, in David Hinton's (2019, xv) words, the way a poem "articulates the emptiness surrounding the words. It is not just open to silence, it articulates silence." Hinton quotes from Lao Tzu's *Tao Te Ching*:

> If you aren't free of yourself
> how will you ever become yourself?
> Give up self-reflection
> and you're soon enlightened.
> Give up self-definition
> and you're soon apparent. (51)

We need to free ourselves of ourselves and empty the preconceptions of our own assumptions in order for a poem to reveal itself to us. This is what I mean by cognitively experiencing a Dickinson poem. A cognitive analysis starts with what the poem is saying and doing, not with interpretation.

1 A Cognitive Approach

Cognitive science research identifies the principles of thought that we all hold in common. All readers are cognitive *readers*. A cognitive *analysis*, by contrast, focuses not on *what* we think but the *way* we think. It involves seeing how a text's prosodic, linguistic, and contextual features reveal cognitive activity at work in both reader and writer. There are two aspects to cognitive analysis: conceptual and affective. A conceptual analysis of poetry describes how meanings arise through the ways images and metaphors are structured through mapping across different domains. An affective analysis describes the ways in which the various forms create the sensory, motor, and emotive feelings that the poem motivates for the reader. Integration of both conceptual and affective analyses reveals the way a poem works. When we read a Dickinson poem, we respond to its rhythms and its tone, we understand its import through our knowledge

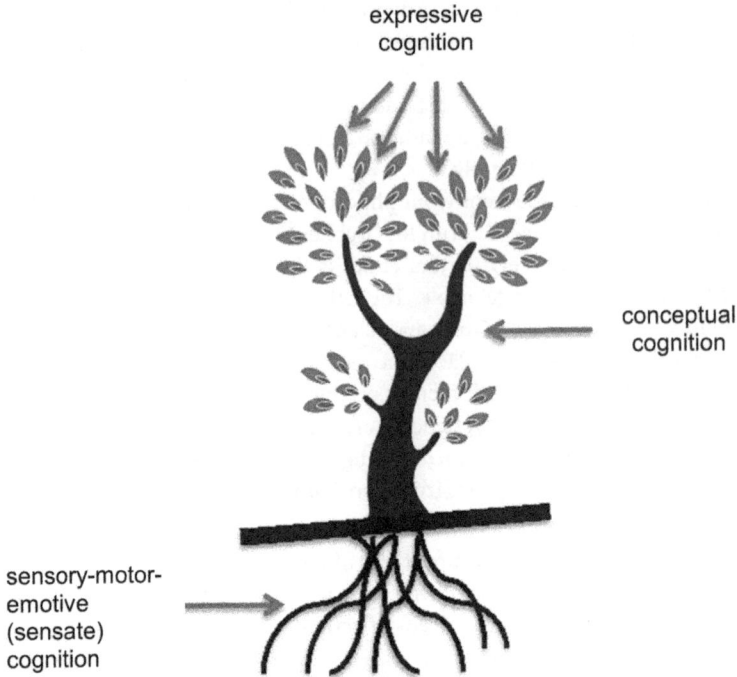

Figure 1.1 The cognitive tree.

both of the English language and the circumstances and contexts of Dickinson's time and place, and we bring our own experience, our own world knowledge, to the text. One advantage of cognitive analysis over purely literary or linguistic readings is that it can show why some readings are compatible and why some conflict (Freeman and McLoughlin 2021).

As living, animate organisms, our own subjectivity is rooted in the integration of the sensory-motor-emotive processes that underlie our conscious, conceptual awareness as we actively interact with the world outside ourselves. The affectiveness of a poem comes primarily from what we hear: the sensory relation of sound to sense (MacLeish 1960). When spoken aloud, a poem's inflections, intonations, pauses, and so on, animate the sensory, motor, and emotive experiences that lead to conceptual awareness and underlie the language of its utterance. As Alice Oswald has noted, poetry "works at the roots of thinking."[5]

Linguistic expression is at the very top of the process of embodied cognition, like the leaves of a tree emerging from the branches and the trunk that are fed from its roots in the earth (Figure 1.1).

Just as the living tree survives by drawing sustenance through its roots, so do all our cognitive activities depend on sensate cognition. And just as the roots of the

---

[5] https://www.brainyquote.com/quotes/alice_oswald_620777

tree are nourished by the quality of the material components of the earth in which they are embedded, so do sensate experiences draw from the natural worlds of our environment.

The poet is motivated from the roots of sensate cognition to conceptualize and then formulate the language of the poem; the reader works downward from the leaves to discover the life of the poem until it reveals its own existence as a living embodied subject in its own right.

While all readings of poetry are of course cognitive in the broadest sense, a specifically cognitive approach does several things. It takes into consideration the presence of all elements in a poem. These include not just word meaning and sentence structure, but its prosodic effects—its macro structure, its rhythms, sounds, repetitive patterning, capitalization, line breaks, and so on. It considers its contextual aspects, both with relation to other poems in a poet's corpus as well as their application to poetic genre, history, culture, and current events. It explores the interrelationships of a poem's images, its metaphors, its affects motivated by subliminal sensory-motor-emotive processes employed by both poet and reader. It goes beyond a simple linguistic analysis by contrasting what is there with what is not to determine their significance within the context of the poem itself. From a cognitive perspective, analyses that focus on specific elements rather than on the gestalt of a poem as a whole can only partially comprehend what it is doing and the way it is working. A cognitive analysis explains the intuitions of readers by showing how and why the poem is doing what it is doing.

2 A Cognitive Primer for Reading a Dickinson Poem

To fully experience and understand a poem, you have to live with it and let it live in you. It is not possible in one reading or even in many to account for how one cognitively digests a poem. Poetry is labyrinthine, not in the sense of a set path with one center, but in the complexity of a semiotic labyrinth that has many paths and many directions.[6] The following primer describes the general steps I take in cognitively reading a Dickinson poem. These steps are not set in concrete. The choice of what steps to take and in what order very much depends on the poem in question. They are rather those one takes on a long journey into the dark unknown, sometimes stumbling, sometimes coming up against dead ends, sometimes losing one's way, but always with faith in finding a center. Nor does the journey have a destination. There is always yet another horizon of possibilities ahead. What cognitive reading does is open up those possibilities. It enables the cognitive reader to see more clearly what the poem is offering in the way of viable interpretations and what readings fail to capture all its workings within the whole: its gestalt, its tone that indicates its overall theme. If the

---

[6] In this sense, a great poem, one that works, is very much the way Lucia Leão (2002, 58) describes hypermedia: as a rhizome-like structure that "can be connected in different directions and from each of its points."

poem "works"—succeeds in achieving what the poet set out to do—a cognitive reader will be able to catch a glimmer of the poet's own cognitive processes in creating the poem and sense to some extent the poet's subliminal intensions and motivations that led to the writing of the poem.

Kang Yanbin (2021), a Dickinson scholar and translator of Dickinson into Chinese, asked me for a cognitive reading of *The daisy follows soft the sun* (H6, F161/J106/M89). The following account is a tentative map for a cognitive reading of Dickinson's poem. It is not complete, because I have only just begun this journey. It indicates directions and routes rather than accounts of progress. But I hope it will help readers see how to begin the cognitive journey for themselves. The steps outlined constitute my reply to her. The order of the steps is what I actually began to think about as I started my journey.

Step 1. In responding to a Dickinson poem, the first thing I do is consult Franklin's (1998) variorum edition to find the location of any existing manuscripts and to see the history of its publication. Now that the images of Dickinson's original manuscripts are available online, we can see a poem in the way it was written, and what variations there may be if different original versions and alternative variants exist. This is important because Dickinson did not herself see her poems into print. All her editors have to make judgments on such elements as line breaks, capitalization, punctuation, and preferred choice of variants. Seeing the original manuscripts also reveals Dickinson's handwriting at different stages of her life. Since Dickinson's handwriting changes over time, a graphological examination can approximate a time period for the poem's creation.[7] Its graphological appearance can also enable the reader to sense the feelings motivating its physical composition (Shapiro 2001).

Since there is only one extant manuscript for the poem in question, my task is simplified by not having to consider all the possible changes and variations that may have occurred in a poem's several versions. The manuscript may be seen at https://www.edickinson.org/editions/1/image_sets/12173848 (Figure 1.2).[8]

Step 2. I then provide a typed transcript of the manuscript, following as closely as I can its format. In my transcript, I have rendered the extremely small horizontal markings as hyphens, and Dickinson's question and exclamation points. Two markings indicate the rising voice on either side of "Sir" and one the falling voice after "decline." The handwriting, regular lines, and exclamation points reveal this poem to be early Dickinson.

---

[7] Unless there is independent evidence for when a manuscript was in existence, such as dates of letters in which they were sent, dates can only ever be approximate, as Thomas H. Johnson warned in the introduction to his variorum edition. Even then, it does not necessarily indicate the date the poem was created.

[8] The story of Dickinson's manuscripts and publication history is a long and complex one. Many of the manuscript poems found after her death in 1886 had been bound by the poet into little packets or booklets. These were referred to as "fascicles" by the first editors, and this term has been retained by subsequent scholars. Dickinson's manuscripts (with a few exceptions) are archived in the Frost library at Amherst College and the Houghton library at Harvard University.

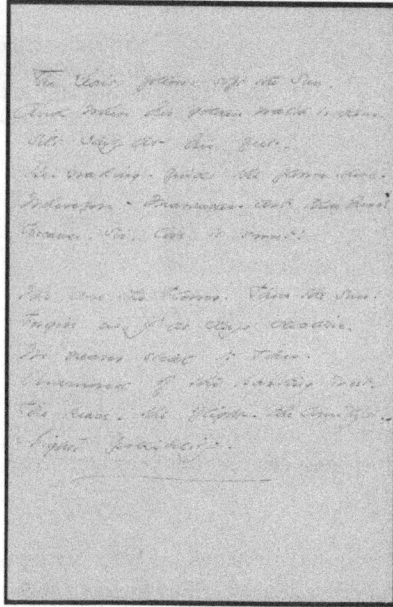

Figure 1.2 *The daisy follows soft the sun* ms_am_1118_3_6_0004, Courtesy Houghton Library, Harvard University.

> The Daisy follows soft the Sun -
> And when his golden walk is done -
> Sits shyly at his feet -
> He - waking - finds the flower there -
> Wherefore - Marauder - art thou here?
> Because , Sir , love is sweet!
>
> We are the Flower - Thou the Sun!
> Forgive us, if as days decline ＼
> We nearer steal to Thee!
> Enamored of the parting west -
> The peace - the flight - the amethyst -
> Night's possibility!

H6, F161/J106/M89

I then print out several copies of the poem on sheets of paper, enlarged and double-spaced, for working on. This is important because I can then look at different characteristics of the poem's prosody, my first main step forward.

Step 3. At this point, I would normally stop and do an intensive study of the rhythms and soundings of the poem, following linguist John Robert (Haj) Ross's (2000)

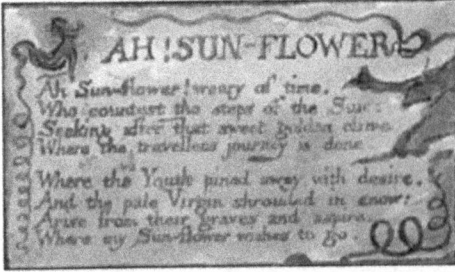

Ah Sun-flower! Weary of time,
Who countest the steps of the Sun:
Seeking after that sweet golden clime
Where the travellers journey is done.

Where the Youth pined away with desire,
And the pale Virgin shrouded in snow:
Arise from their graves and aspire,
Where my Sun-flower wishes to go.

Figure 1.3 William Blake's "AH! SUN-FLOWER."

remarkable practice. These prosodic features of the poem set its tone and thus the feeling it arouses in the reader. I am skipping this step here in the interest of simply outlining my practice rather than delaying this presentation by continuing my actual journey.

Step 4. Because Kang Yanbin had provided me with her essay on the poem, I had already read her review of other literary critics' interpretations. This is a step on my journey that I usually take later. However, because it came early, it already opened several different and, in some cases, conflicting interpretations that have been made. What it did for me was to recognize something not mentioned in Kang's review: How amazingly close Dickinson's poem is to William Blake's in *Songs of Experience* (Figure 1.3).

Both poems have two stanzas. Both have three characters: in Blake they are the sunflower, the sun, and the travelers (youth and virgin); in Dickinson, they are the daisy, the sun, and "We." Both have a similar theme of the sun's passage and the flower's response. As we shall see, there are even closer ties between the events of nature and human life. Both have an opening stanza setting the scene and a closing stanza that complicates in some way that scene. We do not know if Dickinson knew Blake's poem, but the amazing connection between the two suggests she did. Dickinson's poem may then constitute her very different version of Blake's vision.

Step 5. I focus on the poem's prosody, looking first at its macrostructure, and discover an intriguing configuration I had not noticed before.

Each stanza has an equivalent 8-8-6-8-8-6 rhythm and rhyme scheme, the latter reflecting Dickinson's unique rhyming practices. The first three lines and the last three lines of the poem follow the three-part rhyme scheme and comprise in their content a syntactic continuation of the opening scene. They therefore act as a frame for the lines in the middle that comprise a discussion between the sun, the flower, and "We." I also note the chiasmic patterning of repetition in lines 1 and 4 in the first stanza: "Daisy - Sun // He - flower" with a repeat in the second in line 7 that is chiasmic to line 4: "He - flower // Flower - Sun," thus returning to the order of "Daisy - Sun" in the first line.

The Daisy follows soft the Sun -		a
And when his golden walk is done -		a
Sits shyly at his feet -		b
He - waking - finds the flower there -		c
Wherefore - Marauder - art thou here?		c
Because / Sir / love is sweet!		b
We are the Flower - Thou the Sun!		a
Forgive us, if as days decline \		a¹
We nearer steal to Thee!		b¹
Enamored of the parting west -		d
The peace - the flight - the amethyst -		d
Night's possibility!		b¹

Figure 1.4 Structural patterning in *The daisy follows soft the sun*.

Step 6. I raise the question, who are "We"? Even before I made the Blake connection, I had thought, knowing of Dickinson's characteristic style in her two-stanza format of elaborating, complicating, and sometimes controverting in the second stanza what is presented in the first, that the introduction of "We" realizes the incipient metaphor set up by the anthropomorphizing of the sun and the daisy in the first stanza. That is, the poem is setting up a metaphorical blend by which human relations between "We" and "Thou" are expressed through nature's relation of flower and sun. This then raises the question of who "Thou" is. One distinction in the metaphorical mappings that occur happens between the sun's "golden walk" in the first stanza which reflects the daily circuit of the sun and the "days decline" experienced by the "We" of the second stanza. That is, there is a disconnect between the span of a daily occurrence and that of a longer span of years. This reinforces the recognition that the poem is dealing with the metaphor of life's journey, since we are no longer sitting as the daisy is, but "nearer stealing" as the days pass.

Step 7. Other observations include the linking of "Marauder" with "Forgive us." I had already intuited (without having even started step 3 on a complete prosodic phonetic/

syntactic analysis) that the tone of this poem is not angry or supercilious and therefore did not have the same impression this word presents to literary critics who consider the word "Marauder" as being a negative, hostile response made by the sun. I turn to Dickinson's own lexicon, and find the following in Webster's (1844) dictionary:

> MA-RAUD', v.i. [Fr. *maraud*, a rascal; Eth. ᎣᎴᎴ, *marada*, to hurry, to run. The Heb. מרד, to rebel, may be the same word differently applied. Class Mr, No. 22. The Danish has the word in *maroder*, a robber in war, a corsair. So corsair is from L. *cursus, curro*.]
>
> To rove in quest of plunder; to make an excursion for booty; to plunder.
>
> MA-RAUD'ER, n.
>
> A rover in quest of booty or plunder; a plunderer; usually applied to small parties of soldiers.

The word's etymology makes me chuckle, thinking of Dickinson's self-definition as a rebel; also the idea of movement in "hurry, run," and a "rover in quest." I also always turn to the *Oxford English Dictionary* (*OED*), since Dickinson's vocabulary comes as much from her readings as from her lexicon, and find the word *pilfer*. Of course: to steal, a word that occurs in the second stanza. What is the flower stealing? Not a material object, but the sustenance of the sun's rays. What are we stealing? Not a material object, but the movement toward "Thee." This metaphorical transformation of the word *steal* is characteristic of metaphorical blending. Together with "Forgive us" and the connection I find with Blake's poem, I am already inferring the religious as well as the love connotations in this poem.

Step 8. I am now ready to take another detour in this labyrinthine journey of discovery, since a cognitive approach encompasses the historical and cultural contexts surrounding the poem, including all aspects of a poet's and our experiences and knowledge. Blake may have had in mind Ovid's myth of the nymph Clytie and the sun-god in *Metamorphoses*, whereby Clytie is transformed into a flower (the heliotrope) always following the sun. There are many suggestive possibilities that present themselves in considering the myth and Blake's poem and what Dickinson may have taken from them in composing her own poem. This is an important step, because it gives the cognitive reader a more comprehensive picture of Dickinson's conceptual universe based on the schemas of CONTAINMENT and TRANSFORMATION (see Chapters 13 and 14).[9]

Step 9. From here on, my journey has reached a broader avenue and clearer path, allowing me to take on detailed cognitive appraisals, like developing the blending model that will reveal the emergent structural import of the metaphor (see Chapter 9) and exploring further elements I haven't mentioned in this outline, like the affectiveness

---

[9] Cognitive metaphors and schemas are indicated by SMALL CAPITALS to distinguish them from linguistic expressions.

of the final three lines, especially the amazing line "The peace - the flight - the amethyst -," that describe "Night's possibility."

Step 10. I am near the end of my journey of understanding and formulating a cognitive reading of this poem. All these outlined steps would continue to engage me as I follow the cognitive procedure that accords with the classical movements of *oratio*: the level of responding to the spoken word of poetry; *meditatio*: the level of reflecting conceptually on it; and *contemplatio*: the level of experiencing its subliminal affect as we come to understand how the poem reveals itself by what it is doing.

Dickinson's poem can be read on several levels, depending on how one constructs its metaphorical possibilities. Its theme concerns love: of the flower for the sun metaphorically representing "we" and "thou." Although it may be read as a human love story, since "Daisy" was Dickinson's nickname for herself, the fact that the second stanza does not say "I am" but "we are," the plural use of "days decline," the connection between Dickinson's and Blake's poems, the archetypal connection between "the parting west" of sunset and the end of life, and the context of the myth of love for the sun-god, all speak for a metaphorical blending analysis between the Christian journey of life to heaven through love of God.

Linguistic or literary analyses alone are thus not sufficient for understanding a poem's significance. What is needed is a cognitive approach that causes the hearer/reader to participate actively in experiencing a poem. Reading Dickinson's poetry from a cognitive perspective means recognizing what a poem is doing, not simply what it is saying, by focusing on such strategies as

- seeing the entire poem as a *gestalt* on the principle that the whole is more than the sum of its parts;
- recognizing Dickinson's love of philology in the etymological affordances of her word choices;[10]
- considering on what principles readers create analogical and other kinds of mappings;
- exploring the consistent patterns of metaphoring and common themes throughout the poetry;
- exploring the effect of prosodic choices: sound patterns, meter and rhythm, inflections of punctuation that reveal sensory, kinesic, and emotive affects;
- taking into consideration the poet's possible underlying intensions and motivations in order to come closer to Dickinson's own perceptions and feelings;[11]
- establishing the principles of Dickinson's poetics.

That is what this book sets out to do.

---

[10] "Affordance" is the term used to refer to the complementarity of what is offered between the environment and the individual (Gibson 1966, 127).

[11] My use of the term *intension* is deliberate. "Intention" implies conscious purpose in communicating meaning, whereas "intension" (related etymologically to intensity) allows for subliminal, unconscious processes that motivate the poet.

3 Outline of Chapters

This introductory chapter takes up the question of the difficulties Emily Dickinson's poetry presents and outlines the cognitive steps that the ensuing chapters will use in reading her poems. The enigma surrounding her works is compounded by the fact that, except for a few poems, they were not published in print form until after her death.

Since several extant versions of a poem in manuscript occur, and often variant wordings are ˙suggested within a single poem, editors are faced with challenging decisions. It raises the question: "What constitutes a Dickinson poem?" Arguments have arisen in Dickinson scholarship over the importance of reading Dickinson in manuscript form rather than in print editions. Domhnall Mitchell (2005) provides an extensive study of the manuscripts that raises the many questions scholars face. The question then becomes what a reader is reading *for*. Cognitive decisions about the poet's intensions and motivations are based on the version of a poem being read. From a cognitive perspective, therefore, I take into consideration the various possibilities that the manuscripts present in helping us get closer to Dickinson's own cognitive processes as she composed her poems.

Chapter 2, "Everything Counts: Reading the Manuscripts," does not concern itself with critical controversies over manuscript study, but rather focuses on the cognitive insights reading the manuscripts gives in recognizing Dickinson's poetic fluidity. I compare edited versions of *Opon a lilac sea* (A502, F1368/J1377/M590) with its manuscript version to show how editorial rearrangements can obscure the grammar and thus the meaning of a poetic text. I conclude with a cognitive analysis of *Dreams are well* (H183, F449/J450/M225) that clarifies its structure by restoring Dickinson's original line breaks. Paying attention to the physical embodiment of a poetic text can affect the way we respond to and experience it.

Chapter 3, "The Manuscript Markings," is concerned with the physical manifestations of Dickinson's punctuation that affect the way the poems sound. I revisit Edith Wylder's (1971) argument that the various graphological markings (transcribed variously in the edited versions as commas, periods, or dashes of varying lengths) reflect rhetorical inflections of the voice that Dickinson learned from her school elocutionary textbooks. Focusing on inflection, emphasis, and pause, I give a cognitive reading of *It was not death* (A85-7/8, F355/J510/M187) and *All I may* (A91-5/6, F799B/J819/M435) to reveal their sound effects.

Chapter 4, "Measuring Time in Meter and Rhythm," introduces a fresh reading of English metrical form. It focuses on the importance of understanding Dickinson's metrical system in cognitively recreating the rhythmic effects in the breaking of lines in *A sepal, petal, and a thorn* (A82-1/2, F25/J19/M34) and *How sick to wait* (H66, F410/J368/M218).

Chapter 5, "Affective Prosody," takes the reader more deeply into the prosodic elements that create the affective tones and thus the themes of Dickinson's poems. After giving a cognitive reading of *A sloop of amber* (A112, F1599/J1622/M642), I show how attention to the affective prosody of the poem *Of bronze and blaze* (H74, F319/J290/M152) can resolve a critical argument over its interpretation.

Chapter 6, "The Life of Words," takes up Dickinson's love of philology and the potentialities a cognitive approach presents in exploring a word's etymology and the context in which it occurs. First, I look at the poems that directly refer to philology before giving cognitive readings of *A chilly peace* (A97, F1469/J1443/M612) and *We talked with each other* (A516, F1506C/J1473/M621) along with several more poems to show the ways in which Dickinson's word choices bring the poems to life.

Chapter 7, "Bringing a Poem to Life," addresses the ways in which a cognitive analysis enables the reader to follow the labyrinthine affordances offered by an otherwise puzzling poem.

The next two chapters take up the question of Dickinson's grammar. From a cognitive perspective, Dickinson's grammar is not wrong but idiolectic. In Chapter 8, "Intimate Discourse," I show that her grammar conforms to the type of informal, social discourse common among those who know each other well. This explains why it is that readers feel close to the poet even while not fully understanding how her sentences are being structured, or even what they might be saying. Chapter 9, "Grounded -Self Spaces," explores the grammatical question of -self pronoun usage. Neither traditional nor generative grammar can explain Dickinson's use of the -self pronoun. However, by applying the cognitive principles of mental space theory, I show that Dickinson's use of the -self pronoun, while complex, is grammatically consistent throughout her poetry. Dickinson's manipulation of projections from one mental space to another is determined by perspective and grounding of the various selves involved. By using the principle of -self pronoun projection from the subject/agent in one mental space into another, Dickinson creates for us a *world of possibilities*: a world in which things can happen and be made to happen through the agencies of the self.

Chapter 10, "The Presence of Self," then explores the various selves Dickinson adopts with the possibilities of reciprocal, antiphonic relationship, not only with others but with the natural world of animals, plants, weather, death, life, and time itself. In addition, the rise of the dramatic monologue in lyric crept into her poetry as she adopted many different roles and voices.

The next two chapters discuss in more detail cognitive mapping strategies from readers' perspectives. Chapter 11, "The Way We Map," focuses how readers utilize mappings in reaching their interpretations. First, I introduce the principles of different kinds of cognitive mapping strategies that readers employ in general, based on an informal experiment I conducted. I then present an example of readers using mapping strategies in an online discussion of *Of God we ask one favor* (A819, F1675B/J1601/M655) and show how a cognitive blending analysis can help determine which readings may be compatible or preferred.

Chapter 12, "Intentional Mapping: The Search for Coherence" presents three case studies that reveal how readers are applying their own ideological positions on the question of whether there is life after death to the poem *I heard a fly buzz* (A84-1/2, F591A/J465/M270). I then provide a cognitive reading that traces the context of Dickinson's own reading and the forms of feeling in the poem.

Chapter 13, "Conceiving a Universe," outlines the metaphorical and schematic underpinnings of Dickinson's poetics based on her rejection of the linear LIFE IS A

JOURNEY THROUGH TIME metaphor. I first present a cognitive account of Dickinson's complex notion of time before introducing the AIR IS SEA cognitive metaphor that underlies the creation of Dickinson's conceptual universe based on a CYCLE schema and the metaphor LIFE IS A VOYAGE IN SPACE.

Chapter 14, "A Transformative Poetics," considers Dickinson's dominant schemata of CONTAINMENT and TRANSFORMATION that underlie her poetics to show how metaphor is the dominating structure for a poet's poetics in establishing the iconic nature of poetry. First, I explore the notion of CONTAINMENT that triggers metaphors of confinement, and then TRANSFORMATION that triggers metaphors of escape, with both schemas at work in *Dare you see a soul* (A162, F401C/J365/M214). I then discuss the role of metaphor in creating a poem as icon in *When Etna basks and purrs* (H374, F1161/J1146/M705) and its realization in *My cocoon tightens* (H189, F1150/J 1099/M494) to show how iconicity serves to trigger the poem as reaching beyond itself as a transformative experience, both for poet and reader.

In Chapter 15, "Dickinsonian Cognition," I conclude this introduction to a cognitive reading of Dickinson's poetry by raising the question of what it is that makes a Dickinson poem recognizable as such. Arguing that cognitive analysis can identify and authenticate a poet's poetics, I take up the example of how applying the concepts of poetic cognition was able to detect that a manuscript accepted by most Dickinson experts as authentic was in fact a clever forgery. A cognitive approach may also be able to contribute to an explanation of poetic evaluation: what makes a poem work. I end by summarizing the significance of poetic cognition in current literary and cognitive science research.

The various chapters explore some of the features of Dickinson's poetry that are relevant to cognitive analysis. They are not inclusive. It is hoped that, by understanding what it means to take a cognitive approach, the reader will see the kind of strategies to adopt in order to experience what a Dickinson poem is doing in reflecting the poet's experience of "felt life" (Langer 1967, 64):

> [W]ho has a naïve but intimate and expert knowledge of feeling? Who knows what feeling is like? Above all, probably, the people who make its image—artists, whose entire work is the making of forms which express the nature of feeling. Feeling is like the dynamic and rhythmic structures created by artists; artistic form is always the form of felt life, whether of impression, emotion, overt action, thought, dream or even the obscure organic process rising to a high level and going into psychical phase, perhaps acutely, perhaps barely and vaguely.

The questions poetic cognition addresses include the following non-exhaustive list:

1. What elements of literary discourse are common to human reasoning in general and what distinguishes the literary from the non-literary? The methodologies of conceptual metaphor, figure-ground perspective, deixis, prosody, iconicity, semiotics, and stylistics attempt to address this question.

2. What can literary creativity tell us about mind/brain processes and their emotive affects? This question falls under the rubric of neuroscience and psychology.
3. What are the mechanisms that enable creativity to occur and to be recognized and understood? Answers to this question are being attempted by research in neuroscience, cognitive psychology, and conceptual integration networks, among others.
4. What can a cognitive study of literature contribute to the question of what cognitive strategies are universal and what culturally bound? These are the concerns of sociological and anthropological linguistics.
5. What determines the aesthetic qualities of a literary text, and how can or should those qualities be evaluated to distinguish the excellent from the good from the inferior? This introduces the philosophical notions of ethics and evaluation.

I take up several of these cognitive approaches as I present examples of how a cognitive reading can help us both to experience what a poem is doing and illuminate Dickinson's cognitive workings as poet. In spite of common assumptions about the reasons for Dickinson's seclusion from life and her inscrutability, she is, as Thomas John Carlisle (1987, 64) says in his poem "Emily Dickinson": "the woman with the perfect word / demure as dynamite."

Everything Counts

Reading the Manuscripts

In a poem, everything counts. That is, it is not just the words and their meanings that engage us. It is the way words reflect off each other, combining and colliding in a variety of ways that draw us in to seeing patterns and repetitions of sounds, plunging us more deeply into experiencing the physical existence of the words on the page. As we read a poem aloud, we become conscious of the rhythms and inflections of the voice that communicate not meanings but feelings.

Living as we do in a technological age of computerized print, with the teaching of cursive handwriting disappearing from many of our schools, we tend to forget how much regularizing of language into print standardizes the format of letter, line, and punctuation. Spontaneous handwriting, of whatever kind, has "fictive motion" as the writing hand moves across the page. When we type, our hand movements are very different, and in computer inscription, the lettering conforms to the type of font chosen. The idiosyncrasies of handwriting that tend to draw attention to the physical appearance of the words on the page are removed, and the eye can more readily make the language "disappear" as minding conceptually projects the meanings the language generates. Something, therefore, is inevitably changed when script translates to print. And if what is changed is part of the necessary features of the language's design, how we respond to and understand what we read will inevitably be changed too.

Writers who oversee the translation of their work from script to print, or who themselves compose in print form, can, to some extent, control these effects. For a writer like Emily Dickinson, who did not publish her poems in print form, the effect of transforming her script into print can drastically alter the way the reader will conceptually experience her poems. The editor who attempts to render Dickinson's poems into print must be critically aware of the consequences of compromising the physical manifestations of her writing, for it is these physically embodied manifestations that enable us to cognitively conceptualize the range of possibilities generated by a Dickinson text (Werner 1995). These involve not merely the meaning, but also the embodied sensory, motor, and emotive affects embedded in a poem. In the past, readers of Dickinson's poems have tended to overlook Dickinson's manuscripts because they have not recognized that language is embodied, just as minding is embodied.[1]

When we read and respond to a poem, we are using our cognitive faculties. These faculties function to integrate our subliminal sensory perceptions, the motor (kinesic) coordination of our bodies, and our emotive predispositions with our conscious conceptualizations: what is meant by "embodied cognition" (Newen et al. 2018). Reading a Dickinson poem cognitively means experiencing what the poem is saying and doing, before arriving at any interpretation a reader might draw from the poem. This is not a step most students are encouraged to make or that literary critics necessarily take. When we understand a poem cognitively, we are *experiencing* the way it is working. When we *interpret*, we apply other contexts—our historical knowledge, our own life experiences, and so on—to our reading. To read cognitively is not only to see a poem *through* the lens of our cognitive faculties but to see *with* it. To do so means understanding the nature of language as the expression of embodied cognition; that is, not simply recognizing what a linguistic expression *means*, but recognizing how literary expression *affects* us.

Manuscript studies of Dickinson's poetry in recent years have focused more closely on the representation of her poems as they appear in her own handwriting (Hart and Smith 1998; Howe 1987). Such studies have given rise to controversial questions over the poet's conscious intention and a reader's interpretation (Mitchell 2005). Those are not primarily my concern here. Rather, I suggest that access to the manuscripts gives us additional insight into the fluidity of the poetic process.

Dickinson's poems range over a comprehensive and expansive view of the world. She drew not only from her experience of her immediate surroundings but from her extensive readings in daily newspapers and magazines and the books available to her in her family library. She developed in her poems a fluidity of expression in her metrical and grammatical practices. This fluidity is particularly evident both by her line breaks and in the way she freely varied her words and formats in multiple copies of a poem and in the variants to words and lines she indicated in her manuscript copies.

Reading the manuscripts allows us to come closer to the ways in which Dickinson may have been thinking and feeling as she composed her poems.[2] It also gives us insight into the remarkable quality of her poetic expression in capturing and simulating the fleeting moments of her lived reality. Although Dickinson's editors have performed a remarkable service in the care taken to make available Dickinson's poems to the world in print form, their regularization of Dickinson's lines sometimes leads to misreading. In the following sections, I describe in detail two poems in which regularization of the lines obscures the poems' effects.

---

[1] The term *minding* reflects the fact that the word *mind* is a reification—a nominalization of what is actually a process of cognitive activity.

[2] Reading the poems within their fascicle contents, as other scholars have done, opens up a further way of exploring the cognitive effects of grouping individual poems together (Heginbotham 2003; R. Miller 1968; Oberhaus 1995; O'Keefe 1986). Cristanne Miller's (2016) latest edition of the poetry places them within their fascicle order.

1 Understanding the "Frame" of a Dickinson Text

My approach has been to start with the assumption that Dickinson knew exactly what she was doing, that what appears to be erratic in her poetry is so only because we have not fully comprehended the principles of her cognitive grammar. Dickinson has what I will call a "frame grammar," after Charles Fillmore's (1977) phrase "frame semantics."[3] Fillmore explains what he means by the term *frame* as follows:

> I have in mind any system of concepts related in such a way that to understand any one of them you have to understand the whole structure in which it fits; when one of the things in such a structure is introduced into a text, or into a conversation, all of the others are automatically made available. (111)

I have found this true not only of Dickinson's semantics but of her entire grammar. I argue, in fact, that to properly "read" a Dickinson poem, one needs to know her grammar, and that this means not just in the sense of "compositional" grammar as a tool but in the sense of Fillmore's frame semantics and subsequent cognitive theories. That is, to use the analogy Fillmore chooses to explain the difference between the two:

> To know about tools is to know what they look like and what they are made of—the phonology and morphology, so to speak—but it is also to know what people use them for, why people are interested in doing the things that they use them for, and maybe even what kinds of people use them. In this analogy, it is possible to think of a linguistic text, not as a record of "small meanings" which give the interpreter the job of assembling these into a "big meaning" (the meaning of the containing text), but rather as a record of the tools that someone used in carrying out a particular activity. The job of interpreting a text, then, is analogous to the job of figuring out what activity the people had to be engaged in who used the tools in this order. (112)

In other words, adopting a cognitive approach means being aware, not only of the features of the text itself but of the author's tools, tools that may possibly reveal for the reader a glimpse of the motivations and cognitive processes that went into the text's creation.

Subsequent cognitive theory supplemented Fillmore's frame theory by introducing the notion of mental spaces (Fauconnier 1985) and domains Langacker (1987, 1991) that operate as tools in the making of meaning. A domain may be understood to consist of a current "reality space" that includes space, time, and the cultural, social, political knowledge acquired from our lived environment. The term *mental spaces* refer to the dynamic processes we develop as we conceive the possibility of other spaces distinct

---

[3] Fillmore's frame semantics is related to the more recent cognitive linguistic terminology of cognitive-cultural models and event frames (Lakoff and Johnson 1998; Langacker 1987, 1991; Talmy 2000; Ungerer and Schmid 1996).

and distant from our reality space. My approach to reading Dickinson's poems is to recognize both the cognitive frames that structure her poetics and the ways in which she uses those frames—the cognitive-cultural models, or knowledge domains that inform them—and the dynamic use of mental spaces.

In what follows, I trace the cognitive steps I take in order to reach an understanding of what a Dickinson poem in manuscript form is saying and doing. The following poem raises both grammatical and semantic challenges to critical interpretation and, as a consequence, has been little discussed.[4] It first appeared in Todd and Bingham's (1945, 69) edition, *Bolts of Melody: New Poems of Emily Dickinson*, as No. 126:

> Upon a lilac sea
> To toss incessantly
>   His plush alarm,
> Who fleeing from the spring,
> The spring avenging fling
>   To dooms of balm.

Johnson kept Todd and Bingham's line arrangement of the poem in his edition (although his line placements obscure the parallelism their indentation shows), restored Dickinson's capitalization, and assigned it No. 1337. The only changes Franklin (1998) made to Johnson's text were to reflect Dickinson's spelling "Opon" for the first word and to assign the poem No. 1368. In his variorum edition, Franklin also adds below the poem Dickinson's line divisions. Two original manuscripts exist. The complete poem, No. 502 in the Amherst College archives, is a pencil draft on a scrap of stationery. The other, No. L51 in the Houghton archives, is a rendition of the final lines of the poem, beginning "Who fleeing / from the / Spring," that Dickinson sent in a letter as a wedding congratulation to Helen Hunt on the occasion of her marriage to William S. Jackson in 1875 (L444). Dickinson kept the same line divisions in her note to Jackson as they appear in the penciled draft (Figure 2.1).

Many of Dickinson's grammatical and stylistic strategies occur in this short poem. The sea-for-air substitution is a pervasive cognitive metaphor in Dickinson's conceptual universe (see Chapter 13). In this poem, the AIR IS SEA metaphor is expressed as the heavily lilac-scented air of early spring becomes the "Lilac Sea." The practice of reconstructing nouns, whereby their usual meanings are detached from their referents, can be seen in the words "Plush" and "Dooms." But the most difficult strategies of all to comprehend are Dickinson's syntactic moves.

Within one sentence of twenty-three words, Dickinson has created a complex subordinating pattern that can be read in at least two ways. In the printed editions, the word "Alarm" is placed on the same line as "His Plush." Since Dickinson rarely capitalizes verbs, this rendition has led readers to interpret the word as a noun,

---

[4] At a conference on Dickinson in October 1989 held at Amherst College, Richard Sewall, a long-time Dickinson scholar and biographer, pointed out the poem's obscurities and confessed his puzzlement, despite his many years of studying Dickinson, at what the poem might signify.

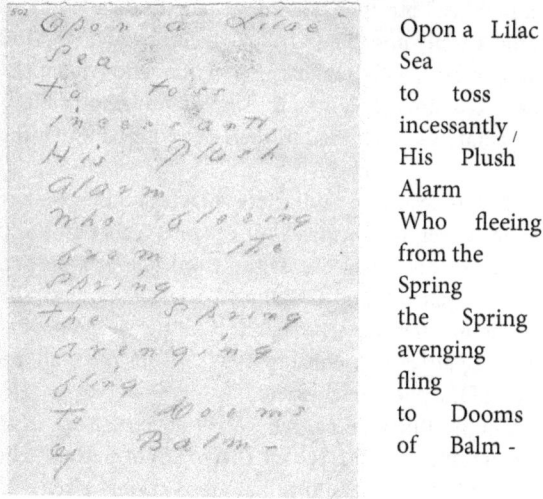

Opon a Lilac
Sea
to   toss
incessantly ,
His   Plush
Alarm
Who   fleeing
from the
Spring
the   Spring
avenging
fling
to   Dooms
of   Balm -

Figure 2.1 *Opon a lilac sea* (A502, F1368/J1337/M590). Manuscript courtesy of Amherst College, Emily Dickinson Collection.

thereby precluding the reading of "Alarm" as a verb with "Plush" as a noun. Recourse to the manuscript resolves this dilemma. Most notably, the word "Alarm" occurs on a separate line, a fact which in itself can account for the capitalization. Although Dickinson occasionally breaks her lines differently when writing various copies or versions of a poem, the fact that she maintains the exact same line breaks in her letter-poem to Helen Hunt Jackson indicates that these line breaks may have had meaning for her.

Understanding the poem grammatically depends on the answer to a simple question: What is the main verb and what the main subject? Dickinson, characteristically, has it two ways. By having divorced the last lines from the extant manuscript version in the letter to Helen Hunt Jackson, Dickinson encourages us to read "fling" as the main verb and "Spring" as its subject. This reading is further encouraged by seeing "His Plush Alarm" as a noun phrase. Reading the poem this way, however, creates difficulties in deciding how to place the complement phrases, and results in an uneasy and unsatisfactory violation of grammar, something I have learned to trust Dickinson not to do.

If one reads "Alarm" as a noun, "Spring/fling" becomes the main subject/verb of the entire poem. Since "fling" is transitive, its object is the preposed noun phrase "His Plush Alarm." The problem created by the reading is a syntactic one: how to fit in the line, "To toss incessantly"? Had the verbal phrase been "tossing," it could be read as an appositive to "His Plush Alarm," and the problem would not exist. But Dickinson did not write this, nor is there any variant to suggest she was unsatisfied with the construction as it stood. "To toss" infers purpose; it is possible to read such purpose as a complement to the main clause: that is, "the spring flings his plush alarm to toss incessantly upon a lilac sea," but then what does one do with "to Dooms of Balm"?[5] Even if it were grammatically possible to have two complements in this context, and one could just manage it here, with total disruption of poetic syntactical order (Spring flings his plush alarm to dooms of balm to toss incessantly upon a lilac sea), the effect is to undermine the final line, since "Balm" does not connote the turbulent image of "toss incessantly."

If, however, "Alarm" is read as a verb, the syntactic problem is resolved. In this reading, "Plush" is no longer an adjective, but a noun. Although the *Oxford English Dictionary* (*OED*) records adjectival use of the word *plush* as far back as 1629, its substantive use is more general, and Dickinson's use of the term elsewhere in the poetry is always as a noun. There are ten references to the word *plush* in the Dickinson corpus, plus one plural *plushes* (Rosenbaum 1964). All of them are nouns, with the ambiguous exception of the poem in question.[6]

There are, according to this reading, five verbal constructions in the poem: *toss, alarm, fleeing, avenging,* and *fling.* Except for the action of fleeing, "His Plush" is the object of all the actions in the poem. After the prepositional phrase "Opon a Lilac / Sea" which begins the sentence, Dickinson employs a subject-object-verb (SOV) order, one that is frequently found throughout her poetry; in SVO order this becomes: "To toss incessantly alarm(s) his plush." The last three lines of the poem are subordinated by means of "who" to the noun phrase "His Plush."

Dickinson's seemingly arbitrary love of preposing complements and apparent disregard for grammatical rules are constrained by the dominating principle of parallelism that structures her poetic form. In this poem, for example, the *aabcccccb* metrical/rhyme scheme brings the *b* phrases "His Plush / Alarm" and "To Dooms / of Balm" into prominent proximity with contrasting ideas of alarm and balm, thus causing the remaining lines to collapse into supporting pairs:

Opon a Lilac	
Sea	*a*
to toss	
incessantly ,	*a*
His Plush	
Alarm	*b*
Who fleeing	*c*
from the	
Spring	*c*
the Spring	*c*
avenging	*c*

---

[5] Note that this rearranging of the syntax suggests that the pronoun "His" refers to "Spring."
[6] The referents of four (not counting the poem under discussion) are indeterminate (A91-11/12, F675/ J401/M418; H215, F684/J457/M331; H327, F1164/J1140/M547; A189PC,6, F1772/ J1738/M686), two refer to feet (H151, F617/J589/M303, H ST12, F1708/J1664/M668), two to the caterpillar (H15, F171/J173/M98, A229, F1523/J1448/M624), and two possibly to a flower in a hummingbird poem (H73, F380*B*/J334/M202) and in a bee poem (H195, F1213/J1224/M503).

fling   *c*
to Dooms
of Balm -   *b*

It is no accident that both of these *b* phrases, *alarm* and *balm*, carry the weight of reader opacity in this poem. The ostensible subject of the first two verbs is missing; the subject of the last two verbs is "Spring," leaving the middle verb, the odd one out of the five, with the subject "His Plush." That is, the poem takes, in its subject/verb structure, the following parallel form:

> ? — toss, alarm
> Plush — flee
> Spring — avenge, fling

The only verb used intransitively is, not uncoincidentally, the verb that has "Plush" as its subject (that is, the verb to flee); all the other verbs are transitive, and their objects are all the "Plush" that is central to the poem.

Still not understood at this stage are the subject of the first two verbs (what "Plush" refers to) and why "Spring" is described as avenging. So the next cognitive stage in understanding the poem involves real-time processing of conceptual integration mapping, the way in which a reader constructs meaning from text. When I first read this poem, I assumed in mapping the metaphors that Dickinson was referring to a bee, possibly with the bee poems that include the lines, "Like Trains of / Cars on Tracks / of Plush" (H195, F1213/J1224/M503) and "is lost in Balms" (H202, F205/ J211/M121), in mind. Her bees are suitors, male, all sexual, in their relationships with their flowers. What other fate, predestined for the bee, does spring bring if not the balm of nectar, the honey of the flower, the lilac? Of course. Beebalm. With her botanical and practical knowledge, Dickinson both knew and grew this herb. It doesn't flower until July–August. Is spring compensating for its own early turbulent-causing euphoria of lilac-laden air by flinging the bee toward its fate: the beebalm of summer? What resonances are here! Spring–summer: anticipation–fulfillment. Dickinson themes, if ever there were.

But that nibble at the soul. Why wasn't the bee satisfied with spring? Or the spring not satisfied with the bee? I knew Dickinson saw "Plush" as a noun, but I needed to explore the cognitive frame in which she uses the term. And that's when I found the caterpillar. Caterpillars in their cocoon state are the unwitting victims of air: they can't do anything except be at the mercy of the wind wherever it takes them. Suddenly the missing subject of the first two verbs falls into place. In their cocoon state, caterpillars can't partake of the lilac. Not like the bee. They are at the mercy of being tossed by the wind. Now "Dooms" takes on new resonance: the fate of the caterpillar to become the butterfly, who can partake. Spring can now be seen as the avenger of the wind's actions by flinging the caterpillar in its apotheosis, the butterfly, into summer. The avenging image now seems to make more sense, with "Spring," the ultimate agent, exacting justice for the disturbance and unease expressed in the first two verbs, "toss" and "Alarm," by flinging the caterpillar into its metamorphosic state of becoming a butterfly. Butterflies, too, partake of nectar. An earlier Dickinson poem shows the caterpillar frame working for my interpretation of this poem, in associating the caterpillar with the words *winds alarm* and *plush* (H15, F171/J173/M98).[7]

As is clear from Johnson's annotation to his edition of the poem, Helen Hunt Jackson returned Dickinson's wedding congratulation with a request for an interpretation. Although we don't have Dickinson's letter of response, she may have provided Jackson with an interpretation. In a subsequent letter, Jackson commented: "Thank you for not being angry with my impudent request for interpretations. I do wish I knew just what 'dooms' you meant, though!" (L444a) Read from a cognitive frame perspective, a quite different construction can be placed on these words. Helen Hunt Jackson was no fool, nor was she naïve. With this her second marriage, she had already experienced the realities of married life. Her question can be read as an "in" joke between the two friends, with a play on the meaning of "Dooms." The emphases of "do," "just," "you," "though," and the exclamation point of Jackson's last sentence all point in this direction.

The appropriateness of Dickinson's wedding congratulation to her female friend takes on new dimensions, a more Victorian expression of fulfillment (caterpillar to butterfly) than the overtly sexual quality of the male bee. It is more characteristic of Dickinson's sensitivity in using appropriate words for particular occasions (Lebow 1999). Understanding a Dickinson poem, I would thus argue, is a matter of recognizing its cognitive-cultural-contextual frame.

2 The Cognitive Import of Dickinson's Line Breaks

Cognitive analysis of another poem, *Dreams are well*, also depends on a significant line break that changes the grammatical reading of a phrase. It appears in the edited versions as an eight-line poem (H183, F449/J450/M225).

David Porter (1981, 103) has commented on this poem as follows:

> Syntax is garbled because of syllable count in poem 450 as well. The poem is a Dickinson allegory of passage into immortality. Her term "Solid Dawn" presents a typical problem of semantics. She wanted it to mean what she said it meant: sufficient dawn, total dawn, permanent dawn. Other defects are caused by the syllable count: it forces the misleading parallel placement of "better," one as a line stop, the other enjambed. The use of "well" as an adjective and "sweeter" where an adverb is called for is gratuitous. The final stanza must be completely rearranged to make an understandable syntactic chain: The surmising robins would never

---

[7] Several poems written later in Dickinson's life can be seen as more complex redactions of themes in earlier longer poems. Compare, for example, *From cocoon forth* (H149, F610/J354/M300) with *My cocoon tightens* (H189, F1107/J1099/M494).

gladden a tree more sweetly than if they were confronting (singing in) a perpetual dawn leading to no day.

With these comments, Porter is clearly being misled by Johnson's printed version, as the original manuscript copy on page 26 shows (Figure 2.2).

All the editors of this poem read the short lines as runovers and regularized the lines to create eight-line poems of two stanzas that conform to the hymn meters that literary critics have identified as the characteristic metrical structure for Dickinson's poetry (Porter 1966).[8] By doing so, however, they have created a poem Dickinson did not write. Johnson's poem is the one Porter read and rightly criticized. Reading, in Blake's ([1810]1972, 611) words, the "Lineaments of the Countenances" of Dickinson's original manuscript conceptually projects a very different poem with very different results.

Note first the word spacing. It has been assumed that Dickinson was indifferent to where her lines break, that the ends of the paper on which she wrote forced her to runover her lines (Mitchell 2005). But there are plenty of examples to show that Dickinson would crush words together into one line if that is what she wanted. The extra and regular spacings between the words in lines 3 and 4 draw attention to the pattern of repetition they make across the page:

If   One   wake   at   Morn -
If   One   wake   at   Midnight ,

Placing "better" at the end of line 4, as the editors do, destroys the symmetry. The effect of seeing "better," "robins," and "Confronting" on separate lines enables us to understand more easily the poem's structure: what the poem is doing.

Because the editors place "Robins" right after "Surmising," readers gloss the phrase as adjective + noun. Placing "Robins" on a separate line separates the word from "Surmising," thus enabling the latter to be read as a gerund and not as a participial adjective. In this reading it is not the robins that are surmising or confronting the dawn, whether it is a real or "Solid" one. Though Dickinson's robins may faint (H178, F982/J919/M452), they do not surmise. They are, on the contrary, the objects of the surmising, and are placed, we shall see, into equivalence with "a Solid Dawn." In Dickinson's conceptual universe, bees get drunk, flowers shout, tigers thirst, and sparrows know how to starve. But nowhere in Dickinson's nature is the pathetic fallacy committed of assuming that beings other than humans can reason.[9]

The embodied shape of a text constitutes its cognitive structure. The comparison at the outset between waking and dreaming gives a clue to Dickinson's constructive

---

[8] Hymn meter is itself derived from ballad meter (Miller 2012, 49–81). See Chapter 4 for a more detailed account of Dickinson's metrical practice.

[9] An exception may be Dickinson's dog Carlo, the only animal in her entire corpus given human logical intelligence (Eberwein 1998, 41). This is not to say that Dickinson did not ascribe embodied cognition to other non-human beings (see Chapter 10).

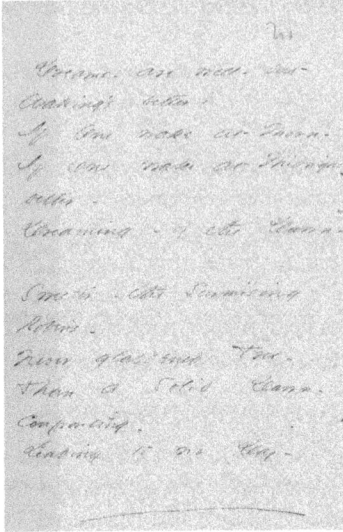

Dreams ˏ are well - but
Waking's better ⁄
If One wake at Morn -
If One wake at Midnight ˏ
better -
Dreaming - of the Dawn -

Sweeter - the Surmising
Robins -
Never gladdened Tree -
Than a Solid Dawn ˏ
Confronting ˏ
Leading to no Day -

Figure 2.2 *Dreams are well* (H183, F449/J450/M225) ms_am_1118_3_183_0004
Courtesy Houghton Library, Harvard University.

design. The poem divides into four equal parts of three lines each, dominated by the four parallel verbs, "Waking" and "Dreaming" in the first stanza, "Surmising" and "Confronting" in the second. The dominant structure of the poem is chiasmic, with words and phrases in *abba* pattern (relevant words in **boldface**):

	Dreams ˏ are well - but
a	**Waking**'s better ⁄
	If One wake at Morn -
	If One wake at Midnight ˏ
	better -
b	**Dreaming** - of the Dawn -
b	Sweeter - the **Surmising**
	Robins -
	Never gladdened Tree -
	Than a Solid Dawn ˏ
a	**Confronting** ˏ
	Leading to no Day -

An analysis of the poem's parallel structure shows that the second stanza is an expansion of the topic and theme of the first, with "Surmising" being to "Dreaming" as "Confronting" is to "Waking." A paraphrase of the poem would thus read: "Waking can only be conceived to be better than dreaming if it occurs in the morning. If it occurs at midnight, it would be better to be dreaming of the dawn. In fact, it is sweeter to surmise/dream that day never occurs at all than it is to confront/wake to a dawn that is solid because it is perpetual."

The poem is thus an argument between what is better: waking/confronting or dreaming/surmising. Waking/confronting is associated with the conscious, logical, and analytical reasonings of human minding, as opposed to the creative, imaginative, and analogical capabilities of dreaming/surmising.[10] The various forms of the punctuation markings in the manuscript text carry the argument forward, as the prosodic markings serve to animate the dialogic voices in the mind.[11] Edith Wylder (1971) observed that Dickinson's markings conformed to the four types of inflections found in *The Rhetorical Reader* used at the Amherst Academy when Dickinson was a schoolgirl there. As its author, Ebenezer Porter (1835) explains, the inflections of the conversational voice support the intended meaning of the utterance; any fault in the use of the correct inflection does violence to the meaning (see Chapter 3). E. Porter distinguishes these inflections by using different notations: when "antithetic relation is expressed or suggested" the rising slide "either expresses negation, or qualified and conditional affirmation" whereas the falling slide "denotes positive affirmation or enunciation of a thought with energy" (44). The monotone, or horizontal dash, "belongs to grave delivery, especially in elevated description, or where emotions of sublimity or reverence are expressed" (27).

Dickinson's markings in the manuscript copy reveal three directions of slant: horizontal, up, and down, with some marks ambiguous between horizontal and down. There is only one upward slant, that after "better" at the end of line 2. The two slants that clearly go down occur at line end, after the capitalized words "Midnight," line 4, and "Confronting," line 11. The eleven remaining slants are all roughly horizontal. No other punctuation appears.

The horizontal inflections reinforce the deliberation of the opening statement: "Dreams - are well -." Immediately, however, another voice is heard, and we as readers are betrayed into the argument of the poem: "but / Waking's better /," and the rising slant not only serves to negate the opening statement but to suggest that the claim is being insinuated rather than strongly expressed, an insinuation that, though conventionally stereotypical in its assertion (waking is better than dreaming), will be undermined throughout the rest of the poem.

This undermining begins with the concessionary third line, as the voice of the poet appears to agree with the second. It is as though the voice that has asserted "Dreams are well -." agrees with its arguer: "Yes, waking is better if you were to wake up in the morning, but if you were to wake up at midnight . . . ," and the falling slide after "Midnight ̣," reinforces the strength of the contradiction as the stereotypical and clichéd thought that "waking is better than dreaming" is ultimately and utterly demolished in the second stanza. Notice that the falling slide after "Confronting ̣," in the second

---

[10] The fifty or so references to dreaming in Dickinson's poetry are all positive. See, in particular, *We dream it is good we are dreaming* (H92, F584/J531/M266).

[11] I am grateful to Jonathan Culpeper for pointing out that Dickinson's inflections look remarkably like those used for marking spoken language in computer corpora. (See Chapter 3 for further discussion of how Dickinson's manuscript markings relate to the creation of sense through sound.)

stanza occurs after the words that parallel the waking at midnight of the first; that is, confronting "a Solid Dawn ‚."

The argument is couched in terms of projections from the present reality space of "Dreams are well -" into hypothetical spaces that contrast dreaming with waking. The first section, lines 1–3, sets out the terms of the argument, with the concessionary "If One wake at Morn -" ending the section. The first line of the following section, "If One wake at Midnight -," begins the parallel and chiasmic *abba* repetitions, repetitions that will dominate the rest of the poem, and introduces the poem's main argument:

a	Waking's better /
b	If one wake at Morn -
b	If one wake at Midnight -
	better -
a	Dreaming -

The associations created by the chiasmic lines "Waking's better / better - / Dreaming -" create a further *abba* repetition:

a	waking
b	better
b	better
a	dreaming

In the second stanza, the complements, "Robins - / Never gladdened Tree -" and "a Solid Dawn - /. . . / Leading to no Day -," of "Surmising" and "Confronting," respectively, are also in semantic as well as syntactic chiasmic equivalence:

a	Surmising
b	Robins (Never gladdened Tree)
	----
b	Solid Dawn (Leading to no Day)
a	Confronting

These parallelisms throughout the poem adhere to a fairly strict pattern of exact equivalences in chiasmic form. As a result, divergences from such exactness are given especial prominence: are "foregrounded," to use Mukarovský's (1970) term. Despite Porter's comment that "the use of 'well' as an adjective and 'sweeter' where an adverb is called for is gratuitous," noting the poem's chiasmic equivalences and parallelisms tell a different story. The poem begins in a different, non-chiasmic *abab* pattern, with "dreams are" and "waking's" forming the *a* components, "well" and "better" the *b*.

a	Dreams are
b	well
a	Waking's
b	better

The repetitions are not in morphological and lexical equivalence:

> Dreams ≠ Waking
> are ≠ 's
> well ≠ better

"Dreams" is a noun, "Waking" a gerund; "well" is an adverb, "better" a comparative adjective. The only main clause indicative verbs in the entire poem are the first two: *are* and the *'s* of "Waking's." As the lexical and morphological variations that occur in the opening *abab* lines are replaced by the more exact equivalences of the *abba* lines, the expectancy of parallelism is reinforced and an alarm goes off. Something is wrong with the initial argument. And something is wrong. We are being asked to accept something at the conventional and clichéd levels of our experience: "Sure, dreaming is fine, but waking is far superior." Only the poem won't let us do it. The "s" of "Dreams" is not exactly equivalent to the "ing" of "Waking," however much we might want them to be. And the words "Dreams" and "well" are not in the same syntactic category as the words "Waking" and "better." So that we end where we begin. Dreams are well. And the poem shows us why.

The only variation from strict parallelism in the second section of the first stanza is the addition of the prepositional phrase, "of the Dawn." The action of dreaming is initially given lexical form in the opening word of the poem: "Dreams." This nominalization obscures the underlying transitive nature of the verb: one dreams about something. When waking is put into parallel opposition, however, the full force of the distinction is made: whereas the act of waking is intransitive (it doesn't make anything happen, one simply wakes), dreaming is transitive, it creates an object, an object in the poem, moreover, that is experienced by waking "at Morn," that is, "the Dawn."

The second, more complex stanza spells the argument out, as the argument of the poem becomes the argument for all who wake, dream, surmise, or confront. Why is dreaming better than waking at midnight? The hint we have been given in the first stanza, that dreaming is more powerful than waking (since it can produce of its own accord what the latter can only experience), is reinforced by the final section of the last stanza, to which it is compared.

a	Dreaming -
b	of the Dawn -
----	
b	a Solid Dawn -
a	Confronting $_\backslash$

The result of waking at midnight is to experience (confront) the kind of dawn that does not lead to day. It is no accident that "Solid" is a non-paralleled isolate in the poem. It is the only capitalized adjective that is not a comparative and that appears in regular noun phrase position. Porter comments on its importance; it does mean what Dickinson wants it to mean. The phrase results from a complex blending of time and space elements from one input space, someone facing an impenetrable wall, and another input space, the onset of morning. In the blend, because it occurs when waking at midnight, this dawn cannot go anywhere; its creative potential is blocked.

Unlike the waking/confronting spaces, the dreaming/surmising spaces are productive. Holding central position between the comparisons governed by the conditional phrase, "If One wake at Midnight," the robin section is grammatically the most complex structure and the linch-pin or "eye" of the entire poem.[12] The robins are the only named agents in the poem and their action is represented by the only tensed causative verb (to make tree become glad). "Tree," like "Robins," has no determiner at all. As "the Dawn" is the complement and thus the product of "Dreaming," "Robins" are the complement and thus the product of "Surmising." Thus, the third section not only stands in syntactic parallelism with the second section but is brought into semantic equivalence as well.

"Dreaming" and "Surmising" both carry equivalent complements. Just as dreaming can create the dawn, surmising can create robins, which themselves can make something happen (gladden tree). In contrast, waking makes nothing happen, and all that is experienced in confronting is a solid dawn that leads to nothing, that makes "no Day." The syntactic-semantic progression of the determiner has reached its chilling conclusion. From the homophoric "the" of "the Dawn" and its disappearance in "Robins" and "Tree," all associated with the dreaming/surmising side of the argument, it surfaces on the waking/confronting side as the determiner "a" in "a Solid Dawn" that anticipates the negative "no" of "no Day."

But the robins, in the end, do not gladden tree.

a	Robins -
b	Never gladdened Tree -
----	
a	a Solid Dawn -
b	Leading to no Day -

As the robins themselves are placed into parallel equivalence in non-chiasmic *abab* pattern with "a Solid Dawn," both as complements of the surmising/confronting opposition, they reflect the ultimate negation and final undermining of the original argument. The real test of whether dreaming or waking is better occurs not with consideration of the good but of the bad, as Hamlet in his famed soliloquy well knew. Better to dream of something bad (like the non-existence of robins) than to confront it (that robins don't in fact exist). Better to anticipate nothingness than to experience it. With such an argument, anticipation must always win out over actualization and is therefore, characteristically for Dickinson, to be preferred.

---

[12] Ross (2000) refers to the phenomenon of "sore-thumbing," the way a word or phrase or sentence that sticks out from the rest of a poem captures the heart of the poem by its difference in sound, structure, or semantic patterns. The Chinese call this the "eye" of the poem.

We end as we begin, in the poem's reality space. We are forced, as readers, to pass judgment on the poet's initial statement, that "Dreams are well." "Well," as a different category word from its comparative adverbial analogs "better" and "sweeter," stands alone. It appears to be the one grammatical oddity of the poem.[13] This statement is not allowed to have a comparative, unlike the other statements in the poem. The reason becomes evident in the restoration of the *abab* pattern at the end. Any possible alternative to dreaming, given the condition of waking at midnight, is to awaken us to that solid dawn that leads to no day: that is, the stasis of death.

Dreaming, therefore, in keeping us cognitively alive, is good for us: it makes us "well." David Porter interprets the poem as being "a Dickinson allegory of passage into immortality." Instead, I read it as celebrating the creative superiority of cognitive power over passive experience that can only react to someone else's creation. The poem is a chilling refusal to accept the conventional and comforting beliefs of unimagining minding, as the poem poses the existential question always uppermost in Dickinson's thoughts: "Where go we - / Go we anywhere / Creation after this?" (A223, F1440/ J1417/M604). It is a poem about life and poetry, a celebration of the life-force of the imagination over the deadening effects of logic and reason.

Read from a cognitive grammar perspective, the two poems discussed reverberate with a love of language, potent with the force of compressed expression. Neither obscure nor ungrammatical, Dickinson's language explodes with power.

As the foregoing discussion indicates, some problems raised by a Dickinson text begin with the transformations that occur from manuscript to print. In addition, the difficulties presented by Dickinson's syntax, by the seemingly referent-less quality of her words, by her aberrant punctuation, are all too obviously clear, even to the most casual of readers. Before we can begin to appreciate the possible multiple levels of meaning in a Dickinson text, we need to avoid the trap of misreading caused by overlooking the "discriminate and particular" marks Blake ([1810]1972, 611) recognizes as significant in poetry and art.

Ultimately, the precept of poetic cognition that understanding is embodied shows that meaning, imagination, and reasoning have a physical basis in our experience of the world. The choice in a painting of oil over watercolor, acrylic over crayon, or the choice in music of the cello over violin, French horn over flute, affect the very way we respond to and understand a work of art. No less is true of literary works in the form of the printed book. Blake exercised control over each "Blur or Mark" in his poetry by controlling the means of publication, etching his own plates, designing his own words and pictures. Dickinson chose not to publish, at least publication in a print medium controlled by others. As a result, critical interpretations of Dickinson's poetry can be misled by the poem's appearance in print; the physical representation of the letters

---

[13] "Well" is used adjectivally when it refers to a satisfactory condition, as in "all is well," or indicating health, as in "Is your mother well?" Although Porter criticizes this use in the poem, Dickinson may be punning on the syntax of the word, putting it in the context of a comparative but at the same time indicating that adjectivally it is foreshadowing and reinforcing the main argument of the poem that dreams are indeed good (healthy) for us.

and words, the gaps, and marks on the manuscript page all contribute to a poem's "meaning."

In applying the principles and methods of cognitive analysis to literature, I am not claiming that it produces insights into Dickinson's poetry that cannot be achieved by any sensitive close reading or other literary critical approaches. Poetic cognition, in this sense, is not just another literary theory that contributes readings resulting from its own particular approach. Rather, by revealing the cognitive processes by which a literary work is created and understood, poetic cognition has explanatory power. It illuminates the conceptual structures of a literary work. It explains how both writer and reader make conceptual projections and mappings that create new meanings. It focuses on process, not product. It is limited by these constraints. It cannot explain why a poet like Dickinson thinks the way she does, nor can it (as yet) describe the novel uniqueness of literary creativity. What it can do, as I have tried to show, is illuminate those imaginative capabilities that enable poetry to happen: the real-time process by which we cognitively reconstruct Dickinson's poems by making conceptual projections from the formal shape of their linguistic characteristics.

3

The Manuscript Markings

In her first letter to Thomas Wentworth Higginson in 1862, enclosing four poems, Dickinson's very first question to him was as follows: "Are you too deeply occupied to say if my Verse is alive?" She followed that with another request: "Should you think it breathed—and had you the leisure to tell me, I should feel quick gratitude –" (L260). The idea of poetry having life and breath reflects Dickinson's emphasis on the oral quality of her poems. For it is the prosodic forms of a poem that carry its affective weight: the pitch and inflection of the voice and the stress rhythms of the poetic line that emotively motivate the poet and engage the reader's own affective response.

Except for the special case of purely visual poetry, poetry is meant to be read aloud. The sound patterns of poetry contribute as much to the poem's meaning as do the other characteristics described in the previous chapter. Although all these characteristics depend to a certain extent on the reader's as well as the writer's cognitive processes, reading aloud is very much a matter of performance and thus more clearly represents the reader's cognitive experience and understanding of the poem. But like the other characteristics, reading aloud is governed by the constraints imposed by the text and can suffer if these constraints are ignored or suppressed. A common example is found when students who have not been given the opportunity to study a poem beforehand are asked to read it aloud in class; their discomfiture is a natural result of their inability to reflect in performance a full cognitive understanding of the poem. I have also experienced readings by elocutionists who tend to dramatize individual words and phrases at the expense of the whole, so that the effect and significance of the poem itself are lost. If there can be good and bad readings of a poem, then there must be some cognitive constraints governing the phonetic and prosodic characteristics of the text. Poetic cognition is faced with the challenge of identifying just what those characteristics are. In this chapter, therefore, I discuss the extent to which the various markings in the manuscripts might indicate Dickinson's sensitivity to the cognitive relation of sound to sense in creating their emotive affects.

1 Making Words Sing

In *The Gift of Song* (2020), choral composer Alice Parker encourages us to listen and hear the living song with our ears and hearts, to understand that singing is not reading notes off a page but rather breathing life into the melodies and rhythms of song through our own minds and bodies. What she says about song is true also for poetry. We create poems from the depths of our feelings, and we recreate the living poem by uttering with our voices, hearts, and bodies the poem on the page.

For those of us familiar with Thomas H. Johnson's and Ralph W. Franklin's (now supplemented by Cristanne Miller's) reading editions, it comes something of a shock to see the poems in their original manuscript forms. In these reading editions, Dickinson's markings, when not clearly regular punctuation, like the comma, question, exclamation, and period, are regularized to horizontal lines, whether by means of the hyphen, en–dash, or em—dash.[1] Their standard-length appearance and placement create an obtrusive pattern in our reading experience. In contrast, experiencing the widely different shapes and sizes of Dickinson's markings and the varied ways they appear on the page in the manuscripts make them much less noticeable and intrusive. We need to see them, not as rigid instructions, but as fleeting indications of possible voicings that make the words and phrases of her poems come alive, to make them sing.

One can understand the early editors' decisions not to reproduce them when moving the poems into print, except where they felt they were needed. Todd and Bingham (1945, ix) summed up their reasoning as follows:

> In some poems, dashes are sprinkled about so lavishly that they give to the page the appearance of a thread on which the phrases are strung. At times, the dashes seem so integral a part of the text that an editor is tempted to perpetuate them, lest without them the words should fall apart.

When that "thread" is regularized in the way the later editions represent them, the phrases do appear to be "strung." If the "dashes" indeed prevent the words from "fall[ing] apart," then there must be some cognitive motivation for their presence. What is missing is cognitive consideration as to the possible reasons why Dickinson made prolific use of such markings.

Edith Wylder (1971) argues that Dickinson's markings reflect the notations adopted in elocutionary textbooks prevalent in the nineteenth century. By 1835, five years after Dickinson's birth, Ebenezer Porter's published textbook is marked as its 220th edition.[2] Elocution—the art of speaking well and effectively—was a recognized and universal

---

[1] When the earliest published editions did include the dash, it was invariably the longest em—dash. Johnson shortened this to the en–dash, presumably to reduce its intrusive effect, and Franklin reduced this even further to the hyphen to lessen their cumulative effect. I use the hyphen for the same reason, though reluctantly, even in cases where the editors have indicated conventional commas when the markings may convey other affects.

[2] According to the Smithsonian, the 1835 edition is "rare" (https://americanhistory.si.edu/collections/search/object/nmah_1888807). It includes an introduction by the author dated 1831 and was published after his death in 1834. In the introduction, Porter explains that the book was an expanded and cheaper version of his much earlier *Analysis of Rhetorical Delivery* and designed for younger students at the request of academy and high-school teachers. Both books had many subsequent publications and are still available in print and ebook.

means of instruction, and Ebenezer Porter's *The Rhetorical Reader* the standard textbook. Porter developed a system of special markings that indicate how a passage of written discourse should be delivered orally. It was used in Amherst Academy, and Dickinson was intimately familiar with the practice of following these markings when reading out loud from the Exercises provided. Once students had internalized the delivery system the markings denoted, they were able to both speak and read fluently and well with appropriate stress, pitch, and pause without the need for such aids. In other words, training students to pay attention to the way the voice establishes meaning through such features as pitch, the movement of rising and falling levels, emphasis, and so on, improves their cognitive abilities to capture the underlying sensory, motor, and emotive impulses that reflect the spirit of the author. Once students have mastered such ways to experience a poem, they can then reach the understanding needed for interpretation of meaning.

Six of Porter's introductory chapters before the Exercises cover the following subjects: articulation, inflections, accent, emphasis, modulation, and gesture. In her book, Wylder focuses on three of these topics, namely inflection, pause, and emphasis, and identifies four unorthodox markings in Dickinson's manuscript poems in addition to the regular three punctuation marks—question, exclamation point, and period—that relate to Porter's notations and that she claims Dickinson used to reflect a poem's oral delivery of meaning.

Wylder's (1971, 3) critics disputed her argument, first broached in 1963, on the grounds that her theory "implies that the poet placed the notations in the lines of her work simply to ensure its proper oral recitation," thus assuming that Dickinson was deliberately marking her poems for instruction on how to read her poems aloud to an audience. This assumption—that elocution was merely for declamation—resulted from the decline in the twentieth century of elocutionary study, and not from its prevalence in the nineteenth century for teaching students how to speak and read effectively.[3] In her defense, Wylder wrote:

> In my reply to Mrs. Ward [the co-editor with Thomas H. Johnson on their editions of Dickinson's poems and letters] I indicated that Emily Dickinson's "synthesis of rhetorical notation and grammatical punctuation" should not be interpreted as "simply directing the reading of her lines 'to an audience'"; rather, the poet "was punctuating in a far more subtle sense than any of her elocutionary tutors or earlier readers would have dreamed" (3).

From a cognitive perspective, Wylder's "subtle sense" takes on specific and revealing significance.[4] Porter himself emphasized the point that careful study was needed to

---

[3] Wylder (1971, 13) notes that "the art of elocution is held in comparatively low esteem today." In class-ridden England, one's speech patterns determine one's status in society. As a child in England, I was sent to an elocution tutor to improve my chances for future success. I am, therefore, sensitive to Dickinson's own elocution experience.

[4] Even as late as 1997, critics were dismissive of Wylder's thesis. For instance, Paul Crumbley (1997, 14) refers to Wylder's work as "notable, though vexed," and sees the markings as indicative of a

study the "sentiment" (the general thought, feeling, or sense) of a passage, "entering as far as possible, into *the spirit of the author*" (vii). I believe that Dickinson had so internalized Porter's notation that she automatically and subconsciously marked her wording as she wrote, not, as Wylder notes, in the "narrow" sense for oral reading but "in terms of Emily Dickinson's own understanding of the importance of tone as the final determinant of meaning in the process of understanding" (68). In saying so, Wylder is, I believe, recognizing the importance of the cognitive principle that experience precedes interpretation.

The regularization of Dickinson's unorthodox markings in printed versions, like the regularization of her lines, thus obscures the markings' cognitive affects arising from the oral quality of her poems. In the following section, therefore, I focus on inflection, emphasis, and pause, as they are indicated through Dickinson's various markings, and provide just a few examples of their emotive affect in experiencing a Dickinson poem.

2 "On the Viewless Wings of Poesy"

It is an old-fashioned word, "poesy," that to the modern ear smacks of sentimentalism. However, it comes from Late Middle English from Old French *poesie*, via Latin from Greek *poēsis*, variant of *poiēsis* "making, poetry," from *poiein* "create." Keats's phrase, "the viewless wings of Poesy" from "Ode to a Nightingale," a bird noted for the sweetness of its song, emphasizes its oral quality. There is no other word that refers so succinctly to the art of making poetry. Dickinson uses it in a poem that she originally sent to a friend. Though the manuscript is now lost, she recorded it in fascicle 39, marking the original wording she used as variants:[5]

> They have a little Odor ╲
> that to me
> Is metre - nay - tis
> ⁺ melody -   + Poesy -
> And spiciest at fading -
> ⁺ indicate   + celebrate -
> A Habit - of a Laureate -

A81-8/9, F505*B*/J785/M403

poetic "voice" with respect to the poetic self's perspective and point of view. His focus is thus on interpretation rather than experience: on how Dickinson's markings lead to a poetics of identity rather than an emotive poetics of Susanne K. Langer's (1953) "felt life."

[5] Mitchell (2005, 23–5) is correct in pointing out that any transcription into print cannot fully capture its manuscript appearance. In my transcriptions, I am not claiming that Dickinson's arrangement of words on the page were consciously intentional. In following them as well as I can I am simply allowing readers to decide for themselves how they affect their own cognitive readings.

Presumably sent with flowers, the poem clearly emphasizes the senses and the equivalence of sound as melody.[6] If we as readers can capture the sounds of Dickinson's melodies, then we can more clearly come to an understanding of her intensions in creating her poems.[7]

In the earliest phase of Dickinson's poetic creativity, as the first eight fascicles show, the poet was conforming to the regularity of lines, with no interline markings, and rarely using rising and falling inflections at line end. Emphasis occurred through traditional punctuation markings (such as the exclamation, the question, and quotation marks), and underlining (transcribed in print editions as italics). Starting with the poems in fascicle 9, interline markings begin to appear more regularly. What were formerly clearly pausal commas are now sometimes rendered to indicate more closely rising or falling slides, with the horizontal markings varying in length and position.

From a graphological point of view, these various markings do not indicate that Dickinson was consciously marking her poems according to the precepts of Porter's Rhetoric. They are neither deliberate, nor consistent in usage. Erratic as they are, however, there are clear indications that she was sometimes subliminally using rhetorical markings for the purposes of vocal emphasis and effecting meaning. The following discussion of Porter's rhetorical system is merely an indication of how sensitivity to the ways in which words and phrases are spoken can help determine the poet's intensions.

2.1 Inflection

Porter identifies various principles of four modifying inflections of the speaking voice, all of which Dickinson uses:

1. *monotone*, marked by a horizontal line "-": the sameness of sound over successive syllables, belonging "to grave delivery, especially in elevated description, or where emotions of sublimity or reverence are expressed" (27): "Stealth's - slow -" (H126, F311*B*/J289/M148).
2. *rising*, turning the voice upward, marked by an upward slant "/": (1) the *pause of suspension*, "denoting that the sense is unfinished" (31): "Like one in danger ⁄ Cautious ⁄ / I offered him a Crumb ⁄" (A85-9/10, F359C/J328/M189); (2) *tender emotion*, such as "grief, compassion, and delicate affection" (32): "He touched me ⁄ so I live" (A85-3/4, F349/J506/M184); and (3) "commonly used at the last pause

---

[6] The related word "posy" both refers to a small bouquet of flowers and a short poem inscribed on a ring or other object. That Dickinson equated self, poem, and flower is well attested in the literature (Farr 1992).

[7] "Intention" implies conscious attempt to communicate a specific meaning. "Intension," on the other hand, indicates preconscious motivation in creating expressive affect. Intension is closely allied to the word *intensity* and the nineteenth-century use of "sentiment" to indicate affective motivation. The use of the word *intention* in Dickinson studies has led to inordinate arguments that do not characterize the subliminal nature of poetic motivation (see Mitchell 2005, 400–2, footnotes 36–42).

but one" in a sentence to create harmonic variety before the natural fall of the voice at the end (33): "Not for the Sorrow ⸝ done me ⸜" (H76, F333/J276/M146).

3. *falling,* turning the voice downward, marked by a downward slant "\": (1) "*authority* of *surprise* and of *distress,*" indicated by the imperative mood, denunciation and reprehension, and exclamation not linked to "tender emotion" or a question (34): "they called it 'God' ⸜" (H109, F292/J293/M137); (2) emphatic *succession* of particulars "to fix attention" on each one (35): "The Horror welcomes her ⸜ / again ⸜" (A85-11, F360/J512/M190); and (3) the final pause (36–37): "Without the Entering ⸜" (H19, F521/J597/M255).

4. *circumflex,* marked by a curvature "ʋ": union of the rising and falling slides, beginning with a downward and ending with an upward tone of voice, and expressing a conditional, hypothetical, or ironical tone (37–8): "For only Gossamer my / Gown ʋ" (H165, F479/J712/M239). Dickinson also uses the reverse circumflex ^: rising followed by falling slide, which Porter (1835) refers to as protracting the expected falling slide on an emphatic word "in drawling manner, from a high note to a low one" (31): "And Space stares ^ all around ^" (A85-7/8, F355/J510/M187).

In addition, when the rising and falling inflections occur together in disjunctive words or phrases, the rising slide occurs first, the falling after, as in Dickinson's lines: "And now ⸝ I'm different / from before ⸝ / As if I breathed superior Air ⸜" (A85-3/4, F349/J506/M184). Often, in antithetic relations, where the antithetic object is not given, the rising slide can also be the circumflex, indicating even more qualified affirmation, as in "He is bĕtter" (though still dangerously ill), as opposed to the simple affirmation, bétter," that he is cured. Many of Dickinson's lines involve the combination of rising followed by falling slides, indicating antithetical stress disjunction, a factor in speech delivery that supports Paul Crumbley's (1997) thesis of the disjunctive faculty in Dickinson's poetics.

Rising and falling slides can differ in intensity. Since Dickinson places her markings after rather than over the words in question, she is able to reflect the degree of pitch intensity by its position, either above or below the word immediately preceding. The only conventional punctuation marks Dickinson uses are the question, exclamation, period, and comma, though what are often rendered as commas in printed versions of the later poems are actually rising or falling inflections.[8]

The use of markings in elocutionary study is thus designed to *improve* a reader's sensitivity to the emotive undercurrents of meaning. Once students have mastered the principles, then they can more readily enter into "the spirit of the author" without the necessity of such guides. If one follows these elocutionary guides in reading Dickinson's poems aloud, one can capture to a certain extent the cognitive feelings being expressed,

---

[8]   For this reason, I have replaced some commas in a poem's printed version in other chapters with hyphens.

and often their implicit meanings.[9] Recognizing Dickinson's characteristic rhythmic movements through such practice can contribute to reading Dickinson's poems cognitively, without the need of markings to guide the voice. The poem, *It was not death* (A85-7, F355/J510/M187) on page 40 shows Dickinson's inflectional practices at work (Figure 3.1).

The first four stanzas conform to a pattern of rising slides indicating suspension of thought within or at ends of line and horizontal monotone marks at line end, as the scene is set to describe what the state of experience was not and yet was like. Their various levels of placement build a growing sense of conditional intensity to the point of being "like Midnight ⌇ / some ⌇" The barely marked falling slide introduces the abrupt change of tone in stanza 5 as it describes what happens at midnight, with the occurrence of the falling slide marking the transformation of the earlier "Frost" to "Grisly frosts ⌇" and the rare occurrences of the reverse (rising–falling) circumflex, three in all here, that emphasizes the strangeness of cessation of everything ticking and space staring.

The final stanza starts once more with the rising slides as the poem returns to the description of what "it" was like. The emphasis provided by the gravity of the horizontal monotone markings after "Chaos, -" "Stopless, -" and "cool -" falls into negation of any hope with the disjunction of the rising slide after "Chance ⌇" followed by the disjunctive *or*, with two falling slides on "spar ⌇" and "Land ⌇" thus reinforcing by voice change the words "without" and "even."[10]

The strange dot-like markings, given their placement raised above line level both visually and grammatically throughout the poem, are not periods, but seem to mark the "Bells" and "Siroccos" as indicators for being "like Midnight ⌇" with the final one occurring at the very end after the word "Despair," presaging the poem's final revelation. If you attempt an oral rendition of the poem following Porter's descriptions of the rising/falling slides and the reverse circumflex, I think you will see their effect.

2.2 Emphasis

Porter defines emphasis as "*a distinctive utterance of words, which are especially significant, with such a degree and kind of stress, as conveys their meaning in the best manner*" (39). He identifies several kinds of emphasis: absolute and antithetic or relative stress, and emphatic inflection, clause, and double (as in "The *young* are slaves to *novelty*, the *old* to *custom*").

Dickinson frequently uses initial capitalization and underlines words and phrases to indicate emphasis, most of which are examples of Porter's definition of absolute

---

[9] A more accurate rendering of Dickinson's markings would recreate these types and positions. However, since manuscript images are now available online, it is only in this chapter that I discuss the actual markings when relevant as accurately as I can. For this reason, I have, with a few exceptions, used the hyphen mark throughout as the least intrusive of markings.

[10] Porter notes the following example: "We must take heed not only to what we sáy, but to what we dò." If one tries to reverse the order to falling/rising, one does violence to the sense.

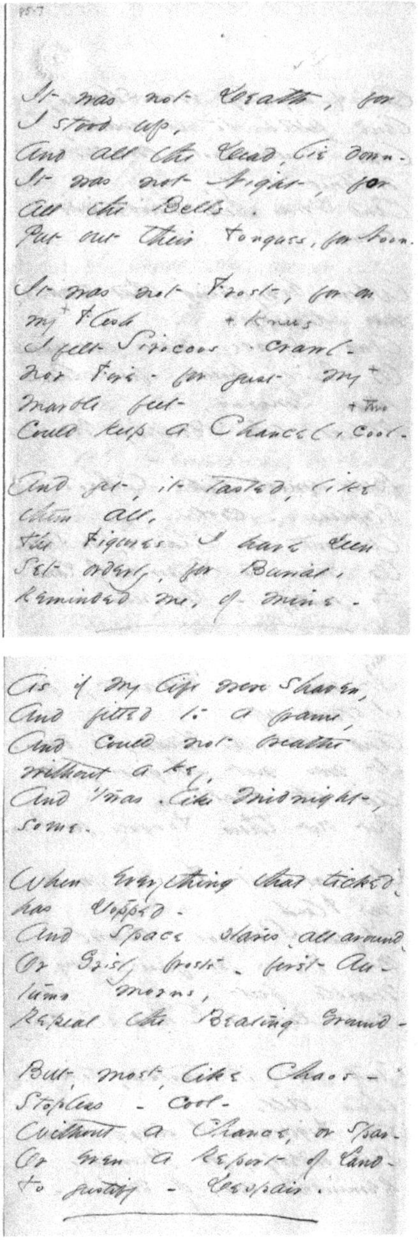

It was not Death , for
I stood up ,
And all the Dead , lie down -
It was not Night , for
All the Bells ·
Put out their Tongues , for Noon -

It was not Frost , for on
my + Flesh + Knees
I felt Siroccos · crawl -
Nor Fire - for just my +
Marble feet   + two
Could keep a Chancel , cool -

And yet , it tasted , like
them all ,
The Figures I have seen
Set orderly , for Burial ,
Reminded me , of mine

As if my Life were shaven ,
And fitted to a frame ,
And could not breathe
without a key ,
And 'twas · like Midnight ,
some ,

When everything that ticked ˄
has stopped - .
And Space stares ˄ all around ˄
Or Grisly frosts , first Aut-
tumn morns ,
Repeal the Beating Ground -

But , most , like Chaos -
Stopless - cool -
Without a Chance , or spar ,
Or even a Report of Land ,
To justify - Despair ·

Figure 3.1 *It was not death* (A85-7, F355/J510/M187) Manuscript courtesy of Amherst College, Emily Dickinson Collection.

stress as a matter of degree. However, in certain cases, and especially with regard to the rising and falling inflections, Dickinson uses stress emphasis as a matter of kind that can affect the sense. This kind of antithetic, or relative, stress emphasis can indicate that the expression is being suggested ironically, hypothetically, or comparatively, all of which are distinctively characteristic of a Dickinsonian tone, as the following example, *All I may* (A91-7, F799B/J819/M435) shows (Figure 3.2).

The poem plays on the contrast between "large" and "small," emphasized by capitalization of "Larger" versus the lack of capitalization of "small ⸝" The rising slides after "All I may ⸝" and "small ⸝" in the first line affirm the hypothetical *if* that the all is indeed small, followed by the question that introduces the antithetical idea of small being larger because of its "Totalness." The next three lines serve as an example of what is less than "Totalness," even in the bestowing of something as great as a "World" when it withholds "a Star."[11]

Such "Economy" is ironically false, because the giving of all, however small, is true "Munificence." Smallness is thus equated with "Utmost," versus the greater, which is "Less." The final lines deliver the irony that "Less ⸝" than "Utmost ⸝," though it may be "Larger ⸝" marked by a rising slide, is "Poor -," marked by the grave monotone of assertion.

2.3 Pause

Competent speakers, Porter notes, will create a natural pause when they "would fix attention on a single word, that stands as immediate nominative to a verb" (58). Pauses are lengthened, he continues, to indicate a "solemn and deliberate call to attention" (59).

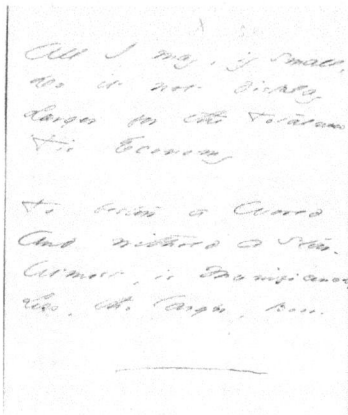

All I may ⸝ if small ⸝
Do it not display
Larger for the Totalness ⸜
'Tis Economy

To bestow a World
And withold a Star ⸜
Utmost ⸝ is Munificence ⸜
Less ⸝ tho' Larger ⸝ Poor -

Figure 3.2 *All I may* (A91-7, F799B/J819/M435) Manuscript Courtesy of Amherst College, Emily Dickinson Collection.

---

[11] Strangely, Franklin does not emend the spelling of "withold" as he is wont to do.

In addition, the *emphatic pause* occurs "sometimes before, but commonly after a striking thought is uttered, which the speaker thus presents to his hearers, as worthy to command assent, and be fixed in the memory, by a moment of uninterrupted reflection" (59).

More than any other rhetorical device at her disposal, Dickinson makes frequent and extensive use of the pause, marked by both her markings and by increasing space between words on the page. In the poem, *It was not death*, quoted earlier, extra space is given between "Siroccos" and "- crawl -," before " - like Midnight ," between "space" and "stares," after "Stopless -," and both after "justify -" and before "Despair," all of which also accompany inflectional emphasis.

Much has been made of whether or not Dickinson chose to break her lines the way she did. In arguing against the notion that they were deliberate, Mitchell (2005) uses measurement of spacing on either side of the manuscript page to give "a better idea as to which spaces are deliberate and which accidental—or at least allows us to entertain the latter possibility" (208). None of these arguments consider that spacing is rather a function of subliminal processes as graphologists have noted (Shapiro 2001).

Dickinson's manuscript markings thus help to relate written forms to their speech sounds. As Alice Oswald notes:

> I've always thought that in poetry, to have no punctuation makes for more punctuation, because it means you really notice the kind of joints and pauses between phrases because you have to. Whereas, I sometimes think that if you put in the punctuation, people will read poems more like novels. They will kind of be searching for the sense rather than hearing the sound of the grammar.[12]

Dickinson's pausal markings and spacings have various functions, both sonically and structurally. But the most dynamic of all are those that are metrically based. The flexibility of Dickinson's metrical style is thus the subject of the next chapter.

---

[12] https://www.poetryfoundation.org/podcasts/154015/a-conversation-with-kit-fan-and-alice
-oswald. Accessed July 7, 2022.

4

Measuring Time in Meter and Rhythm

Dickinson's questions to Higginson in her 1862 letter also reflect her concern that for a poem to be recognized as a poem, it must in some way conform to the genre of poetic form in which she is writing. The punctuation markings in her manuscripts indicate the way her poetry sounds, but they also both reflect and counteract the metrical theory of English poetry dominant in her time. Ebenezer Porter's (1835) *The Rhetorical Reader* is primarily concerned with prose. However, he includes a brief section on *The reading of poetry* (64–6) which must have been of special interest to Dickinson. After stating that "to preserve the metrical flow of versification, and yet not impair the sense, is no easy attainment," Porter provides six general principles, as follows (excluding his examples):

1. In proportion as the sentiment of a passage is elevated, inspiring emotions of dignity or reverence, the voice has less variety of inflection, and is more inclined to the monotone.
2. When the sentiment of a passage is delicate and gentle, especially when it is plaintive, it inclines the voice to the rising inflection; and for this reason, poetry oftener requires the rising inflection than prose.
3. The *rights of emphasis* must be respected in poetry. When the language of a passage is strong and discriminating, or familiarly descriptive or colloquial, the same modifications of voice are required as in prose. The *emphatic stress* and *inflection*, that must be *intensive*, in prose, to express a thought forcibly, are equally necessary in poetry.
4. The *metrical accent* of poetry is subordinate to sense, and to established usage in pronunciation.
5. The *pauses* of a verse should be so managed, if possible, as most fully to exhibit the sense, without sacrificing the harmony of the composition. No good reader can fail to observe the *caesural* pause, occurring after the fourth syllable. There is another poetical pause, occurring at the *end of the line*.
6. The vowels *e* and *o* when apostrophized in poetry, should be preserved in pronunciation. But they should be spoken in a manner so slight and accelerated, as easily to coalesce with the following syllable. As:

> But of the two, less dang'rous is th'offence,
> Who durst defy th'Omnipotence to arms.

Porter's principles reflect an emphasis on pronunciation within metrical versification.

1 Weird and Wonderful Melodies

"Music," Alice Parker (2020, 56) writes, "flows like water, inevitable, adapting to all challenges, but eventually finding its way home to the sea." The motions of music are the motions of poetry. The pace of movement and the suspension of pauses all contribute to the presentation of sense through sound.

As children, the Dickinson sisters had both singing and piano lessons. Emily in particular learned to improvise her own tunes and melodies on the piano, often at night. Kate Scott Anthon, writing to Sue Dickinson in 1917, reminisced about "the old blissful evenings at Austin's! Rare hours, full of merriment, brilliant wit, and inexhaustible laughter, Emily, with her dog, & Lantern! often at the piano playing weird and beautiful melodies, all from her own inspiration, oh! she was a choice spirit!" (Leyda 1960, I: 367). Her familiarity with the principles of composing music and the hymns she sang in church helped to form the measures of her verses.

Cureton (1992) has persuasively established four-beat structure as the pulse underlying English rhythmic verse forms (see also Attridge 2012). The intuitive response that one hears four prominent beats in English metrical verse, and the occurrence of multiple non-stress or lesser-stressed syllables between those beats, even in iambic pentameter, speaks to the flexibility and variation that neither Germanic nor Romance meter has. Dickinson's manipulations of meter work through the stress rhythms of speech. Whereas an English metrical line distinguishes between weak and strong stress, speech rhythm has up to five-stress emphases (Figure 4.1).

Greater focus is placed on the topic, weather, and even greater on the point being made, that it was dry. Rearrangement of the words or stress placement can change focus and import, whether the comment concerns the month or the region, or the import concerns it being unusual. If the greatest stress is placed on the verb *was*, it would denote agreement with a previous comment made. Placing speech rhythm within a given meter constrains to a certain extent the possibilities of articulation, so that it is rhythmical performance that determines the affect and import of poetic phrasing (Tsur and Gafni, in press).

If one vocalizes Dickinson's lines through rhythmic phrasing, what appears unmetrical to the proponents of syllabotonic meter sound perfectly fine. And there is no reason to demand that a poem as a whole should consist of a set number of

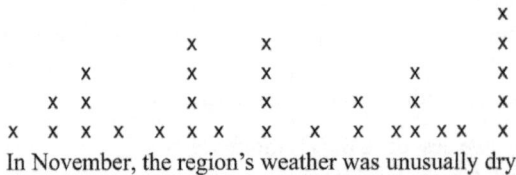

```
                                                      x
                       x       x                      x
              x                x       x              x
              x   x            x       x      x   x   x
          x   x   x   x    x   x   x   x  x   x  x x  x x   x
         In November, the region's weather was unusually dry
```

Figure 4.1 Stress rhythms in speech (from Selkirk 1984).

positions per line. Even within mainstream poetic forms, variation in line length can provide needed emphasis, as both Chaucer and Shakespeare well knew.

Traditional metrical theory relies on the accentual-syllabic (syllabotonic) system—introduced into English verse when French became the language of the English royal courts and government after the Battle of Hastings in 1066—and dominated by the emergence of the iambic pentameter. However, the flexibility of English poetic meter arises from its inheritance of, and emergence from, two distinct metrical systems: Germanic stress-timed and Romance syllabotonic (M. Freeman 2020b).

The attempt to account for English verse structure under alternating syllabic constraints created tension between the Germanic accentual forms evident in Old English poetry, with its pattern of four beats across two half-lines, and the influence of Romance syllabic forms in the language of Middle English. After experiments with various line lengths by earlier poets to accommodate French influence on English verse, it appeared that the iambic pentameter line afforded the most flexible possibilities for rhythmic variety, most evident in the works of Geoffrey Chaucer (Halle and Keyser 1971).

Most studies have assumed that Romance syllabotonic displaced the Germanic accentual stress-timed system of Old English poetry; for instance, Gasparov (1996, 184) notes that, with Chaucer, "English poetry had been given *a new measure that was free from the generic and stylistic traditions of the earlier one*" (my emphasis). As a result, the meter of poets like Thomas Wyatt, John Donne, John Webster, Emily Dickinson, and Gerard Manley Hopkins was considered "rough" if not downright unmetrical. Ben Jonson's (1619) famous comment to William Drummond on Donne's poem is typical: "That Done [sic], for not keeping of accent, deserved hanging" (quoted in R. F. Patterson 1974, 50).

In contrast to the theory that assumes Romance syllabotonic is the underlying structure of English meter, I see English verse as the child of two parents in Indo-European tradition: Romance (of which French syllabics, itself derived from the quantitative measures of classical Greek and Latin, is one example) and Germanic (the alliterative and accentual stress-timed meters of Old English poetry, is another). Although a child inherits features from both parents and their lineage, the child is nevertheless unique in its own right. The challenge, then, is to characterize that uniqueness.[1]

Attridge (2012, 8) identifies a verse form that occurs in many English poems across the centuries and that needs to be identified "as a recognizable metrical genre in its own right."[2] This verse form is dominated by a four-beat structure with the number of offbeat syllables between the beats varying from none to one or two (and possibly more), and is called "dolnik" from Russian prosody that is also found in English and

---

[1] Double-scope blending theory provides a cognitive explanation for how English meter emerges as a new structure from the projection of topology from both Germanic and Romance.

[2] I am grateful to Eva Lilja for drawing my attention to Attridge's article, and to the existence of "dolnik" versification in English poetry.

German prosodic forms (Tarlinskaja 1992). According to Attridge (2012, 6–7), dolnik verse has the following characteristics:

1. the number of syllables varies from line to line; the number of beats per line—four—is unchanging;
2. there is some freedom in the disposition of stressed and unstressed syllables, in contrast to stricter forms that control the number of syllables as well as the beats;
3. the large majority of stressed syllables are felt as beats, and the large majority of unstressed syllables are felt as offbeats (or elements in offbeats);
4. if a syllable that normally does not carry a strong emphasis, like the first syllable of *under*, is treated as a beat, the forceful rhythm encourages the reader to give it some additional weight;
5. beats can be omitted and experienced silently under very particular conditions; offbeats between beats can be omitted with slightly more freedom;
6. only rarely do more than two syllables make up the offbeat between the beats—the norm is to vary between one and two, to produce single and double offbeats;
7. lines can begin and end on a beat or an offbeat;
8. the disposition of the different types of offbeat is such as to enhance the strength of the rhythm;
9. there are no feet.

Attridge notes that "Emily Dickinson is a master of the form" (12).

Dickinson's poetry, along with that of earlier poets like Thomas Wyatt and John Donne, and later poets, like Gerard Manley Hopkins, has been considered metrically complex (if not in some cases downright unmetrical) because it breaks free from the constraints imposed by syllabotonic meter. The affectiveness of emotive intensity in Dickinson's poems lies in the ways in which Dickinson masters the interplay between the syllabotonic and the stress-timed. Consider, for instance, the following poem:

> A sepal - petal - and a thorn
> Opon a common summer's morn -
> A flask of Dew - a Bee or two -
> A breeze - a caper in the trees ＼
> And I'm a Rose!

A82-1/2, F25/J19/M34

The interaction, or better, integration or blending, of Germanic and Romance metrical structure creates the possibilities in the English poetic line for affective response created by rhythmic variation. Compare, for instance, the two systems in the first line. The break in line 1 before the unstress on the third *s* position is not considered significant in syllabotonic scansion, with strong stress falling on three *s* positions in the line:

> /   /   x   /
> A sepal - petal - and a thorn
> w s   w   s w   s w   s

—but becomes particularly significant if the missing stress is supplied in the pause between the two-half line, four-stress system of Germanic meter, in which a "stress rest" marks a missing alliterative word (Creed 1990):[3]

$$/ \qquad / \qquad (/) \, \rule{1cm}{0.4pt} \qquad /$$

A sepal - petal -   and a thorn

w s w   s w   s   w   s

The principle of more than one syllable occupying one metrical position under certain constraints is a feature of syllabotonic meter. The stress rest sets off the phrase "and a thorn," from the preceding words and thus reinforces this aspect of the flower's description as a clue to the final line.

With no identified pauses, line 2 reads as a regular syllabotonic line. Inserted between lines 1 and 3, it foregrounds the stress-timing of these lines. Note what happens if it is placed first:

> Opon a common summer's morn
> A sepal - petal - and a thorn

Not only is the subject of the poem now the common summer's morning and not the emergence of the rose, it obscures the contrast between what happens on a regular day (marked by a line conforming to syllabotonic meter) and the remarkable occasion of a rose's blossoming (marked by a line in stress-timed meter).

Following Germanic stress, the third line, "A flask of Dew - a Bee or two -," falls into two parts of two stresses each, closed by two rhyming words, with the pause marking the break between them. The fourth line, "A breeze - a caper in the trees -," although ostensibly like the third in structure, differs in placement of the pause and varies rhythmically with an anapestic movement in the second half of the line. The pause has the effect of setting off the two parts of the line, so that the missing stress appears to be lacking in the first part and not in position 6 as would be expected in a normal syllabotonic measure (this is reinforced by the rhyme scheme, italicized in the following scansion):

$$/ \qquad / \qquad / \qquad /$$

3 A flask of *Dew* - a Bee or *two* -

$$/ \qquad / \qquad \qquad /$$

4 A *Breeze* - a caper in the *trees* -

---

[3] For example, the opening of *Caedmon's Hymn* starts with a stress rest:

$$(/) \qquad\qquad / \qquad / \qquad\qquad /$$

  Nu scylun hergan hefaenricaes uard

In his performances of *Beowulf*, Creed marked the stress rest with the sound of the replica of the Old English lyre he had made.

If line 4 were to be read simply as an iambic line with stress maxima in two positions and an unstressed syllable in the sixth *s* position, the rhythmic quality of the line would be lost. Thus, although a syllabotonic scansion shows that the line is perfectly regular,

$$
\begin{array}{cccc}
/ & / & \mathrm{x} & / \\
\end{array}
$$
A Breeze - a caper in the trees -
w   s   w s w s   w   s

it does not capture the rhythmic quality of the line as a stress-timed scansion does. In addition, the stress movement of the first line anticipates the stress movement of line 4, thereby increasing expectancy for resolution that occurs with a triumphant conclusion in the last line:

x   /   (x) (/)   x / ___x___   /
A Breeze -   a caper in the trees –

The stress pattern forces the crowding of the last half line in this line, so that speeding up the rhythm after the pause iconically reflects the fast movement indicted by the phrase, "a caper in the trees." The existence of the pause in line 4, at the place where we should expect a repetition of the structure in the preceding line ("of Dew"), demands that the pattern be rhythmically provided.

Line 5 is a half line that provides the missing element, "a Rose," of the pauses in lines 1 and 4, and the metrical effect of stress suspension is to increase the expectancy of the climactic resolution given in the final line. The insertion of the stress rests at the pause seems necessary to characterize its stress-timed metrical pattern, while the syllabotonic influence in English meter enables the position of and suspension of primary stresses to be much more flexible than is possible in Old English Germanic meter.

2 The Line as Metrical Unit

Dickinson's seemingly arbitrary line breaks that can vary from one copy of a poem to another create a problem in determining the line as a metrical unit. Her early editors regularized the poems by conforming them to their underlying structure of the metrical systems of hymnody. This practice has been followed by later editors. In so doing, the results obscure the ways in which Dickinson's practice in employing the flexibility of English meter creates affective response.

In the following poem, Dickinson's editors regularized the line breaks in creating ten-position lines in the first and last stanzas. The manuscript looks like this, with the horizontal markings represented by hyphens:[4]

---

[4] https://www.edickinson.org/editions/2/image_sets/12169901

How sick - to wait - in any
place - but thine -
I knew last night - when
someone tried to twine -
Thinking - perhaps - that　　　　　　　　5
I looked tired ˬ or alone ˬ
Or breaking ˬ almost ˬ with
unspoken pain -

And I turned - ducal -
<u>That</u> right - was thine -　　　　　　　　10
<u>One port</u> - suffices - for a
<u>Brig</u> - like <u>mine</u> -

Our's be the tossing - wild
though the sea -
Rather than a mooring -　　　　　　　　15
unshared by thee -
Our's be the Cargo - <u>unladen</u> - <u>here</u> -
Rather than the "<u>spicy isles</u>" ˄
And thou ˬ not there ˬ

H66, F410/J368/M218

Dickinson's division of the poem into three stanzas creates an outer frame for the central stanza. The first stanza introduces the theme that no-one else can compensate for the absence of a loved one, which, as we shall see, carries the emotive punch of the poem.

The principles of interplay between Germanic four-stress and Romance five-stress (pentameter: ten positions) patterns express the emotions aroused by the absence of the loved one. In the first stanza, the line breaks can understandably be seen as syntactic enjambment (runovers), so that the stanza can be regularized, as the editors do, into ten-position lines. However, if the line breaks are seen as metrical enjambment, the four-stress character of the lines becomes prominent.

By superimposing Germanic on Romance pattern, one can experience the line breaks as including a pause that is a metrical diacritic, like the fermata in music:

```
    /        /          ⌢ /          /
How sick - to wait - in any   place - but thine -

    /        /         ⌢              /        /
I knew last night - when     Someone tried to twine -

    /         /         ⌢             /          /
Thinking - perhaps - that    I looked tired - or alone -

    /        /       ⌢ /          /
Or breaking – almost - with unspoken pain -
```

The stress rhythms indicated are those of the dolnik beat, even in the middle two lines which, under syllabotonic rules, would have five marked stresses.

The middle stanza in its short half lines with two stresses establishes its emotive affect as the central argument of the poem:

/   /
And I turned - ducal –
/   /
<u>That</u> right - was thine –
/ /
<u>One port</u> - suffices - for a
/   /
<u>Brig</u> - like <u>mine</u> -

The stress levels in "<u>That</u> right" and "<u>One port</u>," given Dickinson's emphases, reflect a phrasing that can be understood as the flexibility of dolnik meter, reducing the level of stress on "right" and "suffices."

The final stanza then expands the argument. In this stanza, the other characteristic of Germanic meter emerges: the existence of stress rests. Preference for the loved one is asserted by a typical Dickinsonian contrast (seen also in the "Sweeter" section in *Dreams are well* discussed in Chapter 2): Better to toss on a wild sea with the loved one than to share a safe mooring with someone else. If one considers the first four lines, as regularized by the editors, they form what look like regular syllabotonic ten-position (pentameter) measures. However, both are metrically problematic under the syllabotonic system.

In syllabotonic meter, Halle and Keyser's (1971) stress maximum principle does not allow for a strong stress to fall on a weak position when bounded on both sides by weak syllables.[5] The first line of the final stanza has only nine syllables, ending with a strong stress, so that an optional first empty position would make the line unmetrical by making "the tossing" a stress maximum. Likewise, if the ten syllables of the second occupy the ten positions, the line is unmetrical by the stress maximum on "a Mooring." Both problems are resolved by scanning the line according to Germanic stress principles as follows:

/   /   / ⌢•   /
Our's be the tossing - wild   though the sea –

/   /   ⌢•   /   /
Rather than a Mooring-   unshared by thee -

---

[5] Halle and Keyser's principles do not take into account the influence of the Old English metrical system on English meter. Dudley Hascall (1974) compiled an impressive list of examples from centuries of English iambic pentameter that violate the stress maximum rule.

A similar pattern of contrast occurs in the final lines:

/   /   /   /
Our's be the Cargo -   <u>unladen</u> - <u>here</u> –

/   /   / /
Rather than the "<u>spicy isles</u>" ˄

/   /   /   /
And thou ᵥ not there ᵥ

It is significant that, in the first line, Dickinson squeezed the wording on her page to fit them in, in contrast to the parallel lines, "Our's be the tossing - wild / though the sea -." Unlike the equivalent half line in "Rather than a Mooring" where its two stresses are completed in the following line, the lack of a pause in "Rather than the spicy isles ˄" gives the line a syllabotonic feel in contrast to the strong Germanic stress pattern of the final line. In addition, the emphatic underlining between "<u>unladen</u> - <u>here</u>" and "<u>spicy isles</u>" reinforces the contrast between the reverse circumflex after "<u>spicy isles</u> ˄" and the regular circumflex after "And thou ᵥ not there ᵥ." The circumflex markings after "tired ᵥ or alone ᵥ" and "breaking ᵥ almost ᵥ" in the first stanza that provide qualified affirmation to the "someone" who tried to give comfort are thus emphatically reinforced in the similar markings in the last stanza that give the real reason for the speaker's suffering.

The freedom dolnik verse gives to the otherwise rigid demarcation of conformation to stress and unstressed syllables creates a "promotion" situation where unstressed syllables can take a beat, as in the "And" of the poem's last line, and "demotion," where stressed syllables occur on an offbeat, as in "Someone" in the first stanza. As Attridge notes, "in full-fledged dolnik verse, promotion and demotion are often given vocal realization, *as the language bends to the demands of the imperious rhythm*" (10, n.19; my emphasis).

As this chapter has tried to show, sensitivity to the cognitive effects of Dickinson's manuscript markings establishes that Dickinson's verse does in fact live and "breathe" within the historical contexts of the poetic forms in which she was writing. Arlo Bates (1890), an early reviewer of the first publication of a selection of Dickinson's poems, commenting on what he called violations of the conventional canons of meter and rhythm, dismissed her poetry on "the score of technical imperfection" while declaring that "it is to be judged as if it were a new species of art" (Buckingham 1970, 29). In contrast, through cognitive analysis, a clearer account of Dickinson's relation to the metrical principles shows the poet's mastery in opening up the possibilities of the variety that English verse forms allow through the blending of two distinct metrical systems.

Although conformity to a stricter metricality of the syllabotonic line influenced the regularization of English metrical theory, the example of poets that hear and respond to its Germanic origins created the increased flexibility of the English poetic measure that resonates in all the different forms of English verse structure. Like Donne and

Wyatt before her, Dickinson was sensitive to the stress rhythms of Germanic meter. The following discussion shows Emily Dickinson at her most skillful in manipulating the demands of dolnik meter to create emotive affect.

The editorial regularizations of line breaks in the following poem reveal the relentless rhythm of a galloping horse that underlies its dolnik metrical structure, a rhythm that is disrupted in the second stanza by the double images of space ("Giant long") and time ("Year long") to end with the passionate cry for a "disc" to bridge the distance between the living and the dead. The first and third stanzas are identical in metrical form, while the second and fourth diverge:

> Under the Light - yet under -
> Under the Grass and the Dirt -
> Under the Beetle's Cellar
> Under the Clover's + Root -　　　　　　+ Foot

> Further than Arm could stretch
> Were it Giant long -
> Further than Sunshine could
> Were the Day Year long –

> Over the Light - yet over -
> Over the Arc of the Bird -
> Over the Comet's chimney -
> Over the Cubit's Head –

> Further than Guess can gallop
> Further than Riddle ride -
> Oh for a Disc to the Distance
> Between Ourselves and the Dead!

A92-3/4, F1068/J949/M478

The word repetitions intensify the regular beats of the horse's hooves, with stress predominantly falling on the even positions of the line. The pause before the end phrase of the first lines of the first and third stanzas creates a slight unstable disruption in rhythm. In learning the principles of musical composition on the keyboard, Dickinson would have been familiar with the devil's note, a musical interval that creates a certain instability, giving an unsteady feeling of a sound that needs to be resolved.[6] Daniel Esparza (2018) describes its history:

Medieval musicians and theorists understood harmony in music as being in an allegorical relationship with the divine. Since a chord is normally composed of three notes, the pleasant sound produced by it was thought of as a symbolic

---

[6] I am grateful to Alice Parker and Emily Seelbinder for their help with my musical naivety.

representation of the unity and trinity of God: three different notes producing a single chord, in perfect harmony. But what if one of those three notes were not "harmonic"? That's the "diabolus in musica" right there. Although it was never strictly forbidden, even the great Guido D'Arezzo would elaborate some rules regarding when and how should this harmonic sequence be used.

Esparza further notes: "What might have seemed dissonant for Bach was surely not dissonant for Gershwin, but it is also true that even non-initiated musical listeners can identify a 'dissonant' sounding note, a 'tense' harmony which is not equal to a sound being off-key."

The intrusion of a different rhythm in the second and fourth lines of the second stanza, with stress falling on the odd positions again disrupts the regular movement, and the rhythm of the final two lines of the poem both invoke the negation of the possibility of reaching the dead in either space or time. Dickinson's manipulation of stress pattern in *Under the light* culminates in the final two lines, as the "Oh" sets off the rhythmic pattern found in the second lines of stanzas 1 and 3, making stress fall on odd positions in *dirt*, *bird*, *dist-*, and the final *dead*. The sound patterning of [d] reinforces the irregular fall in its positions in the words: first (*dirt*), last (*bird*), first (*dist-*), and then both first and last on *dead*. The dead are below, above, beyond all human reach on earth.

Despite Arlo Bates's (1890) comment in his review of the first publication of several Dickinson poems that "[h]er ear had certainly not been susceptible of training to the appreciation of form and melody, or it is inconceivable that she should have written as she did," Dickinson's musical knowledge and her more modern approach to the use of dissonance in reflecting her complex feelings about religion, life, and death characterize the complex metrical and rhythmic movements of her verse.

5

Affective Prosody

When Dickinson's early editors regularized her lines to conform to the standard forms of hymn meter and changed her various manuscript markings into regular punctuation, they obscured the dynamic indications of feeling that arise from such rhetorical notations. As a result, readers of the edited editions are faced with the additional difficulty of experiencing the affects that arise from seeing at "first-hand" the poet's own written transcriptions, including the variant wordings suggested and the different versions of multiple copies of a poem. When a poem is particularly puzzling or difficult to comprehend, the fact that images of the original manuscripts are available online can help the reader come closer to the poet's own creative processes. However, I hope that close readings of the edited versions through cognitive analysis can nevertheless contribute to a better understanding of Dickinson's poems and the fundamentals of her poetics.

Every poem demands its own way of vocal reading. The elements of prosodic presentation determine how one experiences the sounds and rhythms of a poem (see Chapters 3 and 4). The term *prosody* refers to those rhetorical elements of versification constituting metrical patterns, rhythms of speech, pitch, stress, and speed of vocal intonations that reflect subliminal sensory, motor, and emotive processes underlying our feelings and motivations. These elements establish the tone and thus the theme of an individual poem.

Our speech patterns betray those feelings and motivations directly. For instance, consider a factual newspaper report that the Red Sox lost a game to the Yankees. The fact is neutral in tone. However, a Red Sox fan would respond by uttering the words "The Red Sox lost" with a sliding downward movement over the word *lost*. A Yankee fan, on the other hand, would do the opposite: "The Red Sox lost!" with an upward rise of glee. A similar effect can be seen in the respondent's answer to the question "So did it happen?" in the following exchange between two people, depending on how the respondent feels about the topic subject "it":

Neutral   Certainly. (matter of fact: equal inflection over each syllable)
Negative   Certainly. (sad: downward pitch slide on last two syllables)
Positive   Certainly! (happy: upward slide on last two syllables)

Note that the exclamation-point punctuation helps to indicate the positive response.

It is no accident that in both examples, a downward inflection corresponds with a negative, or sad response; an upward inflection with a positive, happy one. Other possible reactions, such as being unsure or expressing disbelief, are also communicated by changes in pitch and tone. Voice inflection thus indicates the existence of certain schemas that connect the subliminal, unconscious forces at work in our bodily experience with conceptual awareness, like the orientation schema that underlies such cognitive metaphorical expressions as GOOD IS UP; BAD IS DOWN.[1] Along with sound and rhythmic patterns that directly convey sensory, motor, and emotive feelings, various schemas, such as ORIENTATION, CONTAINER, BALANCE, and so on, give us access to what otherwise cannot be articulated and thus establish the tone and thus the theme of an utterance.

In reading a poem, it is important to be sensitive to the affordances it presents in the way of establishing its tone. These affordances can occur through images that we conventionally associate with certain feelings and that have become archetypal tropes in a culture, such as the contrast between day and night, light and dark, summer and winter, and so on. Although poets make use of these tropes, they nevertheless can complicate them. For instance, in the play, *Romeo and Juliet*, Shakespeare has Romeo utter "Juliet is the sun" (Act 2, Scene 2). Day and light have positive overtones as against dark and night. However, Romeo's utterance is a protest against the coming of the light of dawn that will separate the two lovers. A close reading of the scene and the play as a whole indicates that it is the darkness of night that protects the lovers against the "light" of the Capulet-Montague "day" feud. Although we conventionally look forward to the coming of spring after winter, Dickinson's opening line, *I dreaded that first robin so* (A85-1/2, F347/J348/M183), introduces a poem that presents a very different emotional mood.

Awareness of prosodic affects is thus crucial in establishing the way we understand a poem's tone. By focusing not only on word choice but also on the sound patterns and their structural organization within a poem, we are able to come to some sense of the way the poet is presenting its content and thus experience to a certain extent the poet's attitude toward it. The following sections provide examples of how the sounds, meters, rhythms, and prosodic structure of Dickinson's poetry prime a reader's affective responses.

1 Sound Patterns

Sound associations in English, such as onomatopoeia or sound symbolism iconically impart sensory affectiveness to meaning (Nordquist 2021).[2] Words like *gleam, glow,*

---

[1] Conceptual schemas, like conceptual metaphors, are placed in SMALL CAPITALS to distinguish them from linguistic expressions.

[2] This does not mean that speech sounds directly suggest specific emotions. Rather, as Tsur and Gafni (in press) note, "perceptual and emotional qualities are typically generated after the event, by an interaction between phonetic and semantic features in a specific text."

glisten give the sensation of brightness of light; words like *slide, sling, slither* the sensation of smoothness of movement; words like *whiz, fizz, sizzle* conjure up the sensation of a certain hissing sound. It is not, however, simply the sounds of words that can convey affect; patterns of sounds across a given discourse can iconically represent the affectiveness of its content. Consider, for instance, the following line from an Emily Dickinson poem (H ST1a, F1686/J1687/M662):

> The Possible's slow fuse is lit

The consonants in the noun phrase alternate between voiced and voiceless, ending in the only verb in the poem with the voiceless consonant [t]. Between the first unvoiced [p] and the last unvoiced [t] comes a succession of voiced sounds that coalesce into the blzslfzz of "-ble's slow fuse is," evoking the sizzle of a fuse to a bomb about to explode.[3]

In one poem, its play of sounds reinforces the emotive affect that arises from watching a glorious sunset (Figure 5.1).[4]

It is notable that the adjectives, nouns, and verbs of the poem, with the exception of *amber*, are all comprised of unvoiced consonants. The repetition of sounds on the subject "Sloop" and verb "slips" of the first line captures the smooth disappearance of the sun below the horizon, contrasted by the sharp sounds of [k] [p] [t] of the sun's disappearing effect in the following lines.

The impression the poem leaves on the reader is as much one of sound as it is of meaning, and the sounds of the poem iconically manifest the poem's movement from the sight of the sun setting to the resulting feeling of ecstasy. The dominant sound consonants in the poem are [s] and [p]: [s] occurs in all lines for a total of nine times, [p] only in lines 1–3 six times, with [p] increasing then disappearing as [s] ends up dominating.[5] The sound [l] occurs three times in lines 1 and 3, attached to the two dominant sounds:

1. sl p sl ps
3. pl

---

[3] I am grateful to Haj Ross who drew my attention to the explosive effect of these sound patterns in his (unpublished) analysis of this poem.

[4] The poem is an example of how intricate even a short one can be when it is read closely. Three manuscripts exist. On one manuscript (F1599A) just the last line, "A Woe / of Ecstasy" appears. The second (F1599B) contains the last line, "A Woe of / Ecstasy" as a variant to "The Son of / Ecstasy -." Some of the line breaks are slightly different from the third (F1599C), which may have been addressed to Professor Tuckerman at Amherst College (A 836). It begins with the words, "Please accept a Sunset -."

[5] The ratios and order in each line are (s—p):

    3—2 s p s p s
    1—1 p s
    2—3 s p s p p
    3—0 s s s

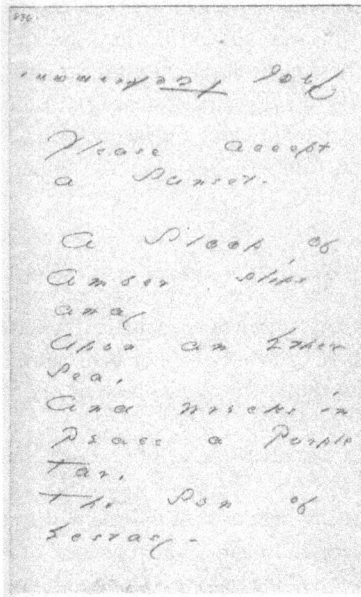

A Sloop of
Amber slips
away
Upon an Ether
Sea -
And wrecks in
Peace a Purple
Tar -
The Son of
Ecstasy -

Figure 5.1 *A sloop of amber* (A836, F1599C/J1622/M642) Manuscript Courtesy of Amherst College, Emily Dickinson Collection.

The change in length of the vowels from *sloop* to *slips* mimics the movement of the sun as it suddenly disappears. The other significant sounds are [k] and [t], both unvoiced plosives, and related to the dominant sounds as follows in lines 3–4:

3. ks t
4. kst s

In the variant, [w] is also significant, occurring twice in the first and variant last lines (see note 2). Without going further and deeper, to put major sounds (excluding vowels) together, the following pattern emerges:

1. sl p sl ps w
2. p s
3. ks p s p pl t
4. s/w kst s

In the variant, the last line picks up the last sounds of the first line with a slight variation of final vowel [wai] to [wo], the [av] from the third sound in the first line, the [ks] and [t] from line 3, and the final [si] sound from the last sound of line 2, capturing/capsizing the sounds of the first three lines into the last. Note that replacing [w] with [s] in the last line doesn't quite have the same effect. My guess is that Dickinson liked the sound-sense patterning of her variant "Woe" with "away," but chose "Son" as having more allure, both in sound and double meaning of Son / Sun. The only consonants left in this version of the last line are [s], [k], and [t], with [k] and [t] between the three occurrences of [s], the last invoking its end rhyme with the second line, "sea." The sound [t] occurs only twice in the entire poem, in *tar* and *ecstasy*. The capsizing still occurs: sksst becomes sksts, as the [t] of "tar" capsizes into the [sts] of the ecstasy of the sea's sunset.

The tar is purple. The color of blood. But of the vein not the artery. Arterial blood is the life-force, venous blood its offspring. The tar is a son of the sea. A son of his ship. The ship slips away. The tar is wrecked. Son of the Sun. Son of God. God of the Sun. What must it have meant to Dickinson to feel the woeful ecstasy of the sunset?

2 Prosodic Significance

Before the onset of literacy, poetry was created and performed orally, so that listeners heard its meters and rhythms. Once oral poetry began to be recorded in written form, respondents encountered it by sight before sound, which they had to reconstruct from their own knowledge of speech rhythm and whatever inflectional clues were provided by punctuation and other markings.

Missing in a linguistic analysis of the written text is the "voice" that is speaking through the words. In speech, it is the tones and inflections, the pauses and the emphases, that communicate the speaker's feelings. The affectiveness of a poem comes primarily from its prosodic form. When spoken aloud, metalinguistic features such as inflections, intonations, and pauses animate the sensory, motor, and emotive feelings that underlie the language of its utterance.

Because such signals are not always communicated by the language itself, written discourse must include strategies that compensate for the lack of oral delivery. These prosodic strategies are both sonic and structural. A cognitive reading of a Dickinson poem takes into consideration the affectiveness of its prosodic elements: its sounds, meter, rhythms, pitch, phrasing, pauses, and so on that create affective response.

Stylistic features can help us identify to some extent the "voice" in a written text, but they cannot, as Stephen Owen (1985, 11) has pointed out, help us to "hear" it. We need, in other words, to listen to the voice behind and emerging from the written word, because it is the voice alone that identifies the speaker. In Owen's words, voice "asks for recognition, points to a person, and engages us in a relation" (109). What gives voice its identificatory cadence are the emotive forces that motivate the speaking. Listening for voice in a poem means experiencing the "minding," the emotive resonance that has produced the text, to hear another voice, not just that of ours imposed upon the text. Dickinson's poems present not only problems in linguistic usage, both syntactic and semantic, but also prosodic elements, such as metrical form, or determining whether line breaks are deliberate or runovers, or contextual allusions that may not be readily apparent. Recreating the voice we hear that is not ours in reading the poem aloud determines how we respond to the rhythms and feel of the poem, and gives us preliminary insight into the poem's feeling and tone. My reading the text aloud determines the interpretation I'm giving to the poem.

The following analysis shows how cognitive methodology can be applied to the complexities and resulting ambiguities of prosodic forms in Dickinson's poetry. Starting with step 1 in my procedural methodology, I choose a poem for analysis, in this case where literary critics disagree. In the poem, *Of bronze and blaze* (H74, F319/J290/M152), disagreement occurs over the use of the pronominal referent "their" in the second stanza.

Applying step 2, determining the text(s), I note there is only one extant manuscript copy of this poem (Figure 5.2).

In my transcript, I have tried as best I could to show the markings and spacings of the original manuscript. There are two alternate word choices in the second stanza. A cross by "An" indicates that it has the variant "some -" that is written at the bottom of the page. "Beetles -" is written directly underneath "Daisies" in the last line. Only one line break differs from printed editions, with "Grass" placed on a separate line. Lines 10–12 present a problem. Line 11 includes the word "manners," which disrupts the rhythm. The customary cross that marks which word is being given an alternate is missing. Looking at the manuscript, "manners" could be the variant for either "Attitudes" or "Oxygen," or an alternate for "stem." (Note that in all these cases there is a variation

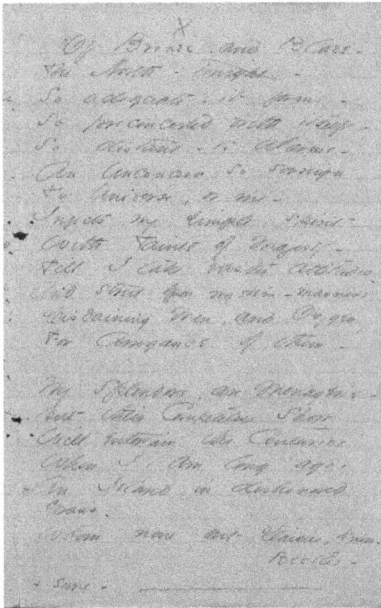

Of   Bronze $_\backslash$ and   Blaze -
The   North -   tonight -
So   adequate -   it   forms -
So preconcerted with   itself -
So   distant   - to alarms -
An   Unconcern   so sovreign
To   Universe $_/$or   me -
Infects my simple   spirit
With   Taints   of   Majesty -
Till   I   take   vaster   Attitudes $_\backslash$
And   strut   opon my stem - manners
Disdaining   Men $_/$ and   Oxygen $_/$
For   Arrogance   of   them -

My   Splendors $_/$ are $_\backslash$ Menagerie -
But   their   Competeless   Show
Will   entertain   the   Centuries
When   I $_/$ am   long ago /
+ An   Island   in   dishonored
   Grass $_\backslash$
Whom   none   but   Daisies $_/$ know -
      Beetles -
+ some -   _____

Figure 5.2 *Of bronze and blaze* (H74, F319/J290/M152) ms_am_1118_3_74_0003 Courtesy Houghton Library, Harvard University.

in syllable count, so that can't indicate which choice might be correct.) The words *manners* and *attitudes* share the same semantic network, and appear to be the most conventional choice, thus probably not Dickinsonian (whoever thought Dickinson conventional?). One could also be said to strut one's manners, so that makes sense too. The alliteration between "strut" and "stem" tends to preclude the *stem–manners* variant choice, though Dickinson's variants are often less vibrant than her first choice. *Men* and *manners* also alliterate, indicating that Dickinson may have thought of "manners" as a variant for "Oxygen." The word *Oxygen* surprises, especially with its last syllable rhyming with "men," and it cognitively fits the attitude adopted by the speaker, as we shall see. All editors have taken the word "manners" as a variant for "Attitudes" and written below it, though if I were to choose it, I would substitute it for "stem":

> Till I take vaster Attitudes ˎ
> And strut opon my stem - manners
> Disdaining Men ˏ and Oxygen ˏ

Jumping to step 3 and linguistic analysis, if the second stanza is taken by itself as an autonomous sentence, the pronominal "their" in line 15 grammatically refers back to the plural subject in line 14: "My Splendors, are Menagerie - / But their Competeless Show." Indeed, many literary critics have read it in just this way. As a result, they claim that the speaker of the poem is Dickinson, and she is referring to her poetry which will survive centuries after her death (Stonum 1990, 144–7). Conversely, other readers see the stanza in relation to the first stanza which describes the aurora borealis and the speaker's response to it, and thus see a stress contrast between "My" and "their" that compares the speaker's "Splendors" with the "Bronze - and Blaze -" of the aurora's "Show" (Vendler 2010, 121–5). Some critics believe the power of the poem lies in maintaining both readings (Hagenbüchle 1986; Salska 1985, 183). Sharon Leiter (2007, 154) notes:

> There are three major elements/realizations in the poem: the grandeur of the northern lights (nature's immortality), the poet's myriad splendors (her poetry's immortality), and the "Island in dishonored grass" (her personal, physical mortality). None of the three predominates. The subtlety and brilliance of the poem lies in the way that Dickinson keeps these realities in motion around one another, letting each reflect upon and modify the impact of the others.

Which of these three approaches most closely reflects the cognitive activity or "minding" that underlies the poem? We may be tempted to think that Dickinson, as she does many times throughout her poetry, is deliberately holding two contradictory ideas at the same time, as Leiter's analysis suggests. However, before we come to any such conclusion, we need to cognitively experience the poem.

So on to step 4 and experiencing the poem's tone, by listening to the voice of the text, hearing the emotional resonances of that speaking voice that is not ours. I listen to the voice of the poem by speaking the poem aloud. I feel an intense disconnect between the language describing the aurora borealis of the Northern Lights and that describing the speaker's response. The aurora is adequate, preconcerted with itself, distant from any noise or other alarms, unconcerned, sovereign. In contrast, the effect on the speaker strikes me as humorous. In line 7, "To Universe ₎ or me -," she promotes herself (I adopt the gender of the poet when no gender for the poetic persona is indicated) into equation with the universe that the aurora disregards; her spirit is simple, it is being infected by the aurora's sovereign unconcern so that it is tainted with feelings of majesty too.[6] The result is that the speaker puffs herself up in an attempt to simulate the features of the aurora. She pretentiously takes on the vaster attitudes that form the aurora's adequacy and self-containment. She egotistically considers herself superior to any other as she struts upon her stem (manners), in opposition to the aurora being distant and preconcerted with itself. The aurora's unconcern with the whole universe causes her to disdain in pure arrogance other men and even the air she needs to breathe. What absurdity!

The second stanza brings her down to earth, indeed, below it, as she recognizes her splendors as only a menagerie, fit for circus performance and entertainment, and imagines her own death, where even the grass is dishonored by her grave, unknown by anyone except the daisies that push up from it or the beetles that share the underground with her. The arrogance collapses as the speaker acknowledges what she had unwittingly described at the outset: that the aurora exists in an absolute, self-contained domain of heaven that removes it from all possibilities of competition or emulation. There is a self-mocking wryness in the speaker's recognition of her own insignificance as she contemplates this truth at the end in the final stanza.

My experience of reading the poem aloud as outlined in step 4 anticipates the cognitive analysis of step 5. Two major domains are being constructed: the heavenly domain of the aurora and the earthly domain of the self. The poem moves from the heights of heaven at the beginning to the depths of the earth at the end. The governing conceptual metaphor is UP IS GOOD; DOWN IS BAD, governed by a VERTICALITY schema. The overreaching self in its strutting attempt to rise above its earthly state topples over into the grave. In addition to these two domains, there are other domains within them that give rise to several conceptual metaphors. The words *infects* and *taints* which Dickinson's Webster's 1844 dictionary defines as interchangeable in meaning, invoke the domains of disease and corruption. The speaker's "Splendors" are metaphorically described as "Menagerie," referring to caged wild animals in traveling exhibitions. That these are conceptual and not merely linguistic metaphors is revealed in the following blending analysis.[7]

The opening four lines create a present reality space ("tonight") describing the speaker's experience of seeing the aurora borealis in the night sky. From that space,

---

[6] It is significant that Dickinson's early editors changed the words "Infects" to "It paints" and "Taints" to "tints," thus downplaying the negativity expressed in Dickinson's original words. They presumably were accepting the grammatically conservative reading of "their" as referring to Dickinson's poetry, and perhaps felt the negative associations conflicted with that idea.

[7] See Chapter 9, section 2 for a more detailed description of blending.

the speaker then projects the two domains: the heavens, containing the aurora, and the earth, containing the self. Through the Vital Relation of cause and effect, the sublimity of the aurora borealis in all its blazing and bronzed glory awakens comparable feelings in the self, which translate into a blended space in which the self puts on her own performance, a "Show," that attempts to emulate the aurora with its own ideas of glory. The speaker in the blend considers herself majestic, above it all, like the aurora itself. However, the language she uses betrays her conceit, her self-deception. Limited by the finite reality of human beings as opposed to the infinite reality of nature, the speaker transforms the various *neutral* descriptions of the "sovereign" aurora, its absolute autonomy of self-containment in its adequacy, its preconcertedness, its distance from any disturbance, its unconcern, into *negative* descriptions of the "majestic" self, its assertions of superiority and power in its overreaching vastness of attitude, its strutting, its disdain, its arrogance. The irony of the poem lies in the fact that although the self thinks that its grandiose attitude in the blend it is creating is being projected from the domain of the heavens, it is actually coming from the earthly domain of disease and corruption.

How does this cognitive analysis that supports the voice I hear in the poem help resolve the linguistic question of which of the three approaches best captures what the poem is doing: "their" as referring to (1) my splendors, (2) the bronze and blaze of the aurora, or (3) both?

If, according to the first interpretation, "My Splendors" refers to Dickinson's poetry, then I think we need to consider the attitude toward that poetry such a reading would implicate. Linking "My Splendors" to the descriptions given in the first stanza makes their overweening pompousness problematic. Even if this link is not made, "My Splendors" are described as "Menagerie," which, as I mentioned earlier would suggest the wild and exotic nature of caged animals in a circus exhibition. Furthermore, we are to believe that the poet in death "dishonors" the grass. Do these descriptions adequately represent Dickinson's attitudes toward her own poetry or poetry in general? I think not. Consider, for example, the poem *I reckon when I count at all*:

> I reckon - When I count
> at all -
> First - Poets - Then the Sun -
> Then Summer - Then the
> Heaven of God -
> And then - the List is done –
>
> But, looking back - the
> First so seems
> To Comprehend the Whole -
> The Others look a needless Show -
> So I write - Poets - All –
>
> Their Summer lasts a solid

Year -
They can afford a Sun
The East - would deem
extravagant -
And if the + Further Heaven -   + Other – final

Be Beautiful as they + prepare   + Disclose
+ For Those who + worship Them -   + to -   + Trust in • ask of -
It is too difficult a Grace -
To justify the Dream -

                                    H137, F533/J569/M292

Here, Dickinson is indeed placing poetry on the utmost pinnacle of creation, above that of nature, or even God's heaven. But note that this is not a straightforward comparison. The sun, summer, and heaven are described as "a needless Show" whereas poets are described by their ability to "Comprehend" them all (a not-so-subtle and characteristic Dickinsonian dig comparing the poet with God). One is reminded of another poem, *The brain is wider than the sky* (A84-5/6, F598/J632/M273). Dickinson's attitude toward poetry in *I reckon when I count at all* is reverential, devotional, not at all the arrogant demeanor of the speaker in *Of bronze and blaze*. In addition, those who accept "their" as referring to "My Splendors" miss the humor of the poem. Consider, for example, another poem on a similar theme, *I send two sunsets*:

I send Two Sunsets -
Day and I - in compe -
tition ran -
I finished Two - and several
Stars -
While He - was making One -

His own was ampler -
but as I
Was saying to a friend -
Mine - is the more
convenient
To Carry in the Hand -

                                    H115, F557/J308/M284

With all its humor, *I send two sunsets* strikes a serious note about the role of poetry to create the semblance of reality. These perspectives on poetry are Dickinsonian, they are not reflected in the perspective of the speaker in *Of bronze and blaze*. For that reason, I would also discount the idea of the third interpretive possibility, that "their" is deliberately ambiguous, referring both to "My Splendors" and the aurora's "Bronze - and Blaze -." Therefore, I conclude that the second interpretation (that "their" refers to the bronze and blaze of the opening line of the poem), being both grammatically possible and cognitively coherent, reflects more accurately Dickinson's own minding in creating the poem. There is irony, however, in the fact that the poet creates a description of the aurora's sublimity in the first stanza that indeed achieves lasting status (Leiter's "immortality") for the poem against its persona who attempts to emulate it.

No one procedure works for encountering and experiencing what a poem is doing. Strategies of reading are as many and varied as the poems themselves. One has to learn to be cognitively attuned to what the poem is offering.

6

The Life of Words

A word is dead, when it is said,
Some say -
I say it just begins to live
That day

A Tr66, F278A.2/J1212/M702

In Dickinson's letter to Higginson quoted in Chapter 3 on whether he thought her verse was alive, she wrote: "The mind is so near itself, it cannot see distinctly, and I have none to ask" (L260). Dickinson's comment reveals her recognition of a word that "just begins to live" can do so when respondents for their part bring it to life. Great poetry captures that quality of illumination: words brought to life as we read and write. In this chapter, I explore two ways in which Dickinson brings her language cognitively to life: recourse to the history of words (etymology), and making words work through stylistic manipulations of language.

Dickinson's vocabulary did not depend only on recourse to her Webster's 1844 dictionary.[1] It developed, like ours, primarily from her experience growing up, the cultural meanings of words in the nineteenth century that were being used around her, and her extensive reading. To fully appreciate the way Dickinson uses words in her poetry, we need to understand the potential meanings they might have in the context of her own experience and thoughts. The advantage of the *Oxford English Dictionary* (*OED*) for us, who live in a different age, is that, unlike other dictionaries, it includes historical meanings of words from the very first time they are found in writing. And because the *OED* includes literary references, some of Dickinson's word associations resonate with them; Shakespeare, for instance, is its most quoted writer.

When we resort to a dictionary, we find, for the words we look up, further meanings we have not acquired through experiential knowledge. Not only do such searches enrich the layers of meaning a word might still have, they can also reveal the semantic evolution of words whose original meanings have been lost. For example, Michael

---

[1] Cynthia Hallen's Lexicon project, online at http://edl.byu.edu/index.php, includes the full text of Webster's 1844 dictionary, as well as lexicon listings for all non-function words appearing in Dickinson's poetry.

Cabot Haley (1988, 40) notes that the word *scruple* from its Latin origins referred to "a small, sharp stone":

> Despite the fact that scruple's quaint history has dropped out of common knowledge, once we are reminded of it, we take conscious pleasure (or displeasure) in finding the worrisome little pebble still there, for a moral scruple is not a major cornerstone of our ethical foundation; it is simply a small pebble of conscience that we seldom think about until it turns up under foot to pang us if we tread on it. The metaphorizing of scruple is an instance of a kind of poetry buried deep in the nature of ordinary language and semeosis. Even dead metaphor fertilizes semantic growth—of language, of poetry, of thought.

Although Haley speaks of dead metaphor, George Lakoff and Mark Johnson (1980) have shown that metaphorical thinking is basic to "the way we think" (Fauconnier and Turner 2002). From a cognitive perspective, metaphor's role in poetry exists on a higher level than the conventional metaphoring Lakoff and Johnson explore. Poetic metaphor brings to life potential meanings "buried deep" in word history. The poetry of Emily Dickinson is an exercise in philology, the love of, or the study of, words—from the Greek *philos*, "loving," plus *logos*, "word."

Three of Dickinson's poems explicitly mention philology: through example, use, and the essential nature of words. The following poem gives an example of philology:

> "Was not" - was
> all the statement -
> The unpretension
> stuns -
> Perhaps - the Comprehension -
> They [+] knew no   [+] <u>wore</u>
> Lexicons -
> But lest our
> speculation
> In inanition die
> "Because God took
> him -" tell us -
> That was Philology -

A505, F1277A/J1342/M566

The quotes in her poem come from Genesis 5:24: "And Enoch walked with God: and he was not; for God took him."[2] Dickinson separates the biblical saying between the two stanzas. The actual saying itself, she says, is simply contained in the phrase "was not." Her spelling of "unpretension" is characteristic of Dickinson's coinage to condense and express in one word a whole range of meaning. Here is Webster's (1844) definition of "pretension":

---

[2]   All biblical references are from the King James Bible, the version Dickinson's family owned and read.

1. Claim, true or false; a holding out the appearance of right or possession of a thing, with a view to make others believe what is not real, or what, if true, is not yet known or admitted. A man may make pretensions to rights which he run [*sic*] not maintain; he may make pretensions to skill which he does not possess; and he may make pretensions to skill or acquirements which he really possesses, but which he is not known to possess. Hence we speak of ill-founded pretensions and well-founded pretensions.
2. Claim to something to be obtained, or a desire to obtain something, manifested by words or actions. Any citizen may have pretensions to the honor of representing the state in the senate or house of representatives. The commons demand that the consulship should lie in common to the pretensions of any Roman—Swift. Men indulge those opinions and practices that favor their pretensions—L'Estrange.
3. Fictitious appearance; a Latin phrase, not now used. This was but an invention and pretension given out by the Spaniards—Bacon.

By prefixing *pretension* with *un-*, Dickinson makes the notion of "claim" disappear altogether. The unpretentiousness of "Was not," its simplicity that we find hard to comprehend, lies in the fact that those who heard the phrase were stunned because they understood directly from their own scriptural experience that to see God is to die: "And he said, Thou canst not see my face: for there shall no man see me, and live" (Exodus 33:20). If Enoch walked with God, he had died; they needed "no Lexicons" to tell them so. Dickinson's variant *wore* (underlined in the manuscript) for "knew" was adopted in another copy she made of the poem. It is another characteristic Dickinsonian play with words: *wore*, physically as in clothing, and *wore* as the past tense of the verb meaning to affect by degrees, to use.

The term *lexicon* comes from the Greek *legein*, "to speak," *dictionary* from Latin *dicere*, "to say." Most of the words we know and use we have never looked up in a dictionary. We acquire our first words through hearing spoken language. We develop their meanings from the context in which we learned them. Most word-meaning networks we accumulate come not from dictionary knowledge but from an encyclopedic acquisition of experience. As a result, the shades of meaning we give to words are more influenced by our own contextual hearing and reading experience than they are from a dictionary.

Unlike those who need no lexicons to understand the statement "Was not," we need more than what is given in that phrase. Dickinson's second stanza moves to our own response in attempting to comprehend it. The word *inanition* comes from the Latin *inanire*, "to make empty, to make void." Inanition is a kind of exhaustion from lack of nourishment, but it also can mean lack of mental or spiritual vitality and enthusiasm. To prevent inanition in our speculation, it is necessary to "mention" the reason: "because God took him."[3] If we no longer know our scripture, then we need

---

[3]   In another manuscript (A 95-10/11), the variant "mention" is given for "tell us."

to be told how and why Enoch "was not," just as we need to resort to dictionaries to understand words we have not otherwise experienced. Dickinson draws on this simple lesson to show the importance of philology. The study and love of the etymological and historical essence of words is needed for words to come alive, to be vital if they are not to die.

Even a seemingly simple word like the verb *take*, occupying seventeen pages in the *OED*, can reverberate in the hands of a poet like Dickinson. Originally having the meaning to grasp, grip, seize, take hold of, the "telling" of God's taking of Enoch as an example of philology becomes in the following poem the idea of searching for words, for the right word, or the word that the poet wants to use, in metaphorical terms of a word being a candidate for a position:

> Shall I take thee, the
> Poet said
> To the propounded word?
> Be stationed with the
> Candidates
> Till I have + finer   + further • vainer
> tried -
>
> The Poet searched
> Philology
> And + was about to ring   + just • when
> For the suspended
> Candidate
> There + came unsummoned   + Advanced
> in -
> That portion of the Vision
> The Word applied to
> fill
> Not unto nomination
> The Cherubim reveal -

A341, F1243/J1126/M557

The metaphor of a word seeking a position in a poem is not simply that of a person seeking employment. The words *nomination* and *candidate* conjure up the realms of politics and religion. Nomination is also related to the linguistic term *nominalization*, with its root meaning of "naming": to turn a verb or an adverb or into a noun, thus reifying action into *res*, Latin for *thing*. The "Word" becomes an active agent rather than a nominated candidate as it seeks to become the most appropriate word for the poet to use. There is an echo of Dickinson's comment in her letter to Higginson of the mind being "so near itself that it cannot see distinctly" in the reference to the word applying to fill "that portion of the Vision." The word vision suggests not simply physical but spiritual seeing, a concept reinforced by the introduction of "Cherubim" in the last line. For Dickinson, such word plays in searching philology raise the level of discourse into the animating spirit of creativity.

In the following poem we have the whole idea of coming to life—the vivacity, the vitality, of the word made flesh. Like the first philology poem, Dickinson also quotes from the Bible, this time with the play on "partook" as a "taking":

> A Word made Flesh is seldom
> And tremblingly partook
> Nor then perhaps reported
> But have I not mistook
> Each one of us has tasted
> With ecstasies of stealth
> The very food debated
> To our specific strength -
>
> A Word that breathes distinctly
> Has not the power to die
> Cohesive as the Spirit
> It may expire if He -
>
> "Made Flesh and dwelt among us"
> Could condescension be
> Like this consent of Language
> This loved Philology.

> H ST14c-d, F1715/J1651/M671

The quotation is from the beginning of St. John's gospel: "In the beginning was the Word, and the Word was with God, and the Word was God." John is referring to the very first chapter of Genesis: "In the beginning, God created the heaven and the earth. And the earth was without form, and void; and darkness was on the face of the deep. And the spirit of God moved upon the face of the waters. And God said Let there be light: and there was light." The power of the spoken word to create light and therefore life is paramount. The opening of St. John's gospel continues:

> The same was in the beginning with God. All things were made by him; and without him was not any thing made that was made. In him was life; and the life was the light of men. And the light shines in darkness; and the darkness overcame it not. . . . And the Word was made flesh, and dwelt among us, and we beheld his glory, the glory as of the only begotten of the Father, full of grace and truth.

In the Genesis passage, we have the beginning, the creation of something. We have the earth without form, empty and dark. The five elements to the process of God creating—spirit, movement, seeing, speaking, and light—participate in making and giving form to earth. They are, I suggest, the same elements that occur in giving form to poetry.

The word poetry itself is from the Greek word *poiesis*, "to make." St. John invokes the Genesis chapter in order to equate God's speaking with the Logos, the divine word that brings everything into being, and the Logos with Christ.

Dickinson plays with the etymologies of words that link the biblical passages with words in the poem in a semantic network of linked meanings. First of all is spirit. The word comes from the Latin *spiritus*, which means breath, and *spirare*, "to breathe," reflecting both noun and verb. To "animate" means to give life to, from the Latin *animus*, meaning both "mind" and "soul." So, to animate is to breathe into, to instill with life. Then there is the Latin word for life, *vita*, the source of our words *vital* and *vitality*. In Latin, *vitalis* is the animating principal, the "life-force" of being and existence. The word *genesis* comes from the Greek word meaning origin, creation, formed from the root *gen-* of *gignesthai*, to come into being, to be born. *Logos*, as mentioned, is the Greek both for word and reason. It is as if Dickinson were anticipating the cognitive revolution with its recognition of embodied minding (Lakoff and Johnson 1980). Words are embodied; they are made flesh (M. Freeman 2002c). When they become so, when they "breathe distinctly," they are "full of grace and truth." I think this is what Dickinson knew above all other things in her "loved Philology."

One reason why Dickinson's poems present such a challenge to her readers is the way she uses words that seem to hover at the edge of our understanding. As readers, we construct meaning from various triggers that are presented: discourse context, the structure of a given expression, and our own experiential knowledge, derived from both the codes and canons of language. Norman Holland (1988, 101) distinguishes between *codes*, "the rules governing letters of the alphabet, numbers, grammar, recognizing a given word as that word, in general, rules that are absolutely fixed for all the people in a given culture," and *canons*, which "express politics or values or beliefs, a person's 'philosophy' in the loose sense, a mental 'set'" or "the intellectual climate of an era." He further distinguishes *background canons* which "reflect heritage, education, and life experiences" from *viewpoint canons*, which "reflect opinions and beliefs" and are thus easier to change (104).

When meeting an Emily Dickinson poem, a reader is faced with a plethora of puzzles. Only by responding directly to one's experience of those puzzles can one come to a full understanding. That does not mean that a sensitive reader does not intuitively grasp the truth of the experience the poem is offering. When we intuitively grasp the gist of a poem, we can (if we are so inclined) work toward understanding why we have that intuition. We notice what the poem is doing and then we arrive at an understanding that both reflects any preliminary vague intuition we might have had and explains why we had it. We can do this best if we pursue Dickinson through the way she makes her words work.[4]

Readers' insights can differ, and in the following analyses I don't want to prejudge what others read by presenting an intuition of my own. The following poem with its many variants encourages us to engage actively in responding to its words:

---

[4] I am grateful to the members of the Emily Dickinson Reading Circle for their contributions to the analyses of the following poems.

A ⁺ chilly Peace infests   + lonesome - • warning -
the Grass
The Sun respectful
lies -
Not any Trance
of industry
⁺ These shadows   + The
scrutinize -

Whose Allies go
no more ⁺ astray   + abroad -
For ⁺ service or   + Honor - • welcome
for Glee -
⁺ But all mankind   + though
⁺ deliver here   + cruise softly here • row [softly here]
                     • sail [softly here] • do anchor [here] -
From whatsoever Sea -

A97, F1469/J1443/M612

First, I note certain places in the poem where I expect one word but get another. The first appears in line 1: I expect "infects" and get "infests." The *OED* tells me that this is a characteristic mistake people often make between the two words. "Infest" in contemporary usage carries the idea of bugs invading a house, and indeed the *OED* has the meaning in verb 2 "to trouble (a country or a place) with hostile attacks; to visit persistently or in large numbers to destroy or plunder." But recourse to the dictionary leads me to other possibilities. Dickinson's Latin and German studies, however meager they may have been (and we don't really have a good sense of that), gave her an orientation to etymology; and her readings, which, unlike ours of the twenty-first century, were extensive both in ancient and contemporary authors, may have enriched her network of meanings for a particular word. So that when I read in the *OED* that "infest" etymologically carries with it the notion of "to fasten, to fix in something," that reverberates for me with Dickinson's image of peace "infesting the grass." Then, I see that an obsolete substantive (noun) usage of *infest* refers to funeral offerings or expiations. Suddenly, what intuitions I had about this poem—why the peace is chilly/lonesome/warning—are deepened and enriched.

Another place I expect one word and get another is in line 5. I expect "trace" and get "Trance." I know what "trance" means for me: a state of suspended consciousness, a dreamlike state. But I go to the *OED*. I discover it is etymologically related to the word transit, a "passage through," and suddenly I have the introduction of movement, which "trace" doesn't automatically give me. Both as a substantive and a verb, *trance* can mean to move about actively, to skip/a skip, to dance/a dance, the idea of rapidity. I note that in Dickinson's poem, such implied movement is negated: "Not any Trance." Then I realize (and this is making words work), that *trance* in its modern sense also involves movement: an entering into a state, not just being in the state itself. Then, as in the word *infest*, I go even deeper: *trance* as a substantive also once meant the suspension of consciousness, the "passage from life to death." Could Dickinson have known that? I think again of her lexicon, her loved philology, the extent of her reading. My preliminary intuitions about this poem (which I have deliberately not shared) seem now to surface willy-nilly, in spite of myself. Or is it that I have recognized these connections because of those intuitions?

Next I look at the structure of the two stanzas. I see immediately a contrast on several levels: the first stanza is land-bound, the second ventures out to sea. The poem is deictically grounded; that is, words like *come, here, this* point to where the speaker is situated, as opposed to their counterparts, *go, there, that*. There is a clear distinction between the "here" of the land and the "there" of the sea, reinforced by Dickinson's choice of "These" (not "Those") shadows in the first stanza. Its subject-agents are abstract or natural: "Peace infests," "Sun respectful lies," "shadows scrutinize." I notice that though "Peace" ostensibly is agent operating on the "grass," grass is actually its topic. From a cognitive viewpoint, *peace infests the grass* is related to the adjective-noun phrase *peaceful grass*. So you have grass, sun, and shadows—all aspects of the natural world. In contrast, the ostensible subjects of the second stanza are "Allies" and "all mankind," both referencing human agents, and possibly other living beings.[5]

I notice too a parallelism between the two stanzas, with "Not any" in line 5 and "no more" in line 10. These negatives introduce the concept of contrast, between what is and what isn't. So when I arrive at the "But" in line 13, I am prepared for a contradiction. The variant has "though." *But* implies contradiction, a statement of "instead." *Though* implies a concessive, naming an exception to a preceding statement. This brings us to a critical puzzle in this poem: What is the antecedent of "Whose"? A syntactic analysis of the previous lines reveals an ambiguity, one typical of Dickinson. Since English is an uninflected language except in its pronoun system (i.e., it does not indicate noun case), the simple unambiguous order of an English sentence is Subject-Verb-Object (SVO). But Dickinson typically goes against such rigid SVO conformity in English syntax. She often preposes the object to the beginning of the phrase, as in "The Wicks they stimulate" in *The poets light but lamps* (A91-13/14, F930/J883/M436). But in a case where the context does not readily resolve the issue, one raises the question, is it an OSV or an SOV structure?

When the participants are animate and inanimate, we intuitively prefer the animate as subject. That is, given the sentence, "Harry drank the coffee" (SVO), either order, "The coffee Harry drank" (OSV) or "Harry the coffee drank" (SOV) will result in an unambiguous reading. But when the action of the verb can be attributed to either of the noun phrases, then ambiguity arises. In lines 5–8, doubly reinforced by the fact that whatever action is being described is negated, is it the case that (no) trance of industry is scrutinizing the shadows, or the shadows that are scrutinizing (no) trance of industry? Either way, the question of the antecedent for "Whose Allies" in line 9 remains. Is it trance, industry, or shadows?

In the transitive use of the word *deliver*, the construction of the second stanza seems straightforward. The allies (whoever they are) no longer are going out there

---

[5] Dickinson referred to her dog, Carlo, as "my shaggy ally" (L280).

but delivering mankind here. With "though," the emphasis switches to a focus on movement: the allies are no longer moving, in going out there, though they are moving, in delivering mankind here. I'm not very happy. It comes back once more to the unresolved questions of a possible referent for "Allies" and the antecedent for "Whose." If one accepts the idea that the allies belong to trance or the shadows, then a certain gothic quality invades the poem, such as the ghosts of the dead straying abroad to ensnare mankind. If the allies are the allies of industry, then I expect some product of trade (silks from China, coffee from the plantations), but instead I get what results from performing the work: service, honor, welcome, glee. I note that "all" in "all mankind" falls on a stress position in the iambic line. Could it be that the allies of industry represent just some portion of mankind that engage in these activities, but that all mankind end up in the same place? Now light glimmers. I go back to my preference for "all mankind" as agent-subject. Is it possible that the allies are in fact part of all mankind?

The question is further complicated, not clarified, by the verb in the conjunctive *but* clause with all its variants. The variants for "deliver" are all intransitive, which makes their agent-subjects "all mankind." But "deliver" is different. In its transitive form, with Dickinson's habit of preposing objects, it would seem that "all mankind" is what is being delivered by "Allies," unless "all mankind" is indeed the agent-subject, and what is delivered is the "service" (with its variants, *honor/welcome*) or "Glee." But I find this a stretch, so again I go to the *OED*, wondering if there is some intransitive use of the verb *deliver*. I find an intransitive use that is possible but not likely, referring to speech or singing, both having a self-reflexive quality (the utterance of words or notes). Self-reflexive use is common, though in the examples cited always accompanying its pronoun ("I deliver myself into thy hands"). And then I find the following usage of *deliver* in pottery and foundries, both intransitive and reflexive: "to free itself from the mould; to leave the mould easily." The earliest reference is 1782 (Wedgwood, Phil. Trans. LXXII. 310): "To make the clay deliver easily, it will be necessary to oil the mould." Could Dickinson have known that? My intuitive sense that *deliver* can be construed as a middle verb form is here encouraged, especially with the semantic networks of meanings associated with material clay and material body.[6] Although the variants for "deliver" clarify the sense of movement toward—*cruise/row/sail/anchor*—and I especially like "anchor" with its additional idea of coming to rest, in the end I prefer the word Dickinson originally chose. The ambiguity heals: either *but* or *though* work, with both their contrastive and concessive meanings, and *deliver*, in this construing, is

---

[6] The middle voice construction is Greek in origin (Bybee 1985, 20–1):

The voice of a verb indicates the role that its grammatical subject plays in relation to the action or state of being expressed by the verb. The middle voice is used mainly to imply that the subject benefits or suffers directly from the action expressed by the verb. It is often the case, though not always, that the subject also represents the cause of that action. In most languages a single argument cannot represent two different semantic roles such as AGENT and PATIENT. In Ancient Greek, however, the middle voice makes this possible. (http://www .greeklanguage.com/grammar/20.html) There is evidence for the existence of a middle voice in English.

a middle voice verb, so that indeed it is all mankind that "deliver" themselves here. This is indeed the height of the poetic capability of making words work.

Where does that leave us? The poem's regular iambic cadence creates no tension, no surprise. Rather a sense of inevitability and calm invades. The deictic grounding of the poem is *here* not *there*, and *here* denotes the land, the natural, peaceful world of grass and sun and shadows to which we come in death, as opposed to going abroad, engaged in human work on the sea of life, a metaphor hovering behind that second stanza with all its variants. Although I expected the variant *abroad* (etymologically meaning "away from the road," i.e., across the sea), I get "astray." The word *astray* carries with it a range of meanings, all denoting the sense of movement away from the right path, a movement into error. I think again about those images of industry, men's purposes in life: service, honor, welcome, glee. All come to the same place, the "chilly Peace" of the grave. And then other early Dickinson poems linking grass to grave come to mind: "the color of the grave is green" (H141, F424/J411/M168); "they perished in the seamless grass" (H140, F545/J409/M298); "an island in dishonored grass" (H74, F319/J290/M152). I note that the noonday sun casts no shadow; the "respectful" sun only leaves shadows when it is low on the horizon. The movement into a suspension of consciousness indicated by the word *trance* reinforces the idea that these are evening shadows, invoking the metaphor that structures the phrase "the evening of life." And yet Dickinson once more surprises. Though one would expect the conventional DEATH IS DEPARTURE cognitive metaphor, we get instead DEATH IS ARRIVING HOME/TO HARBOR, here, where arriving is right and proper, as opposed to the going away/astray, there. Even if it is "a chilly Peace."

The following poem shows complexities of a different kind. It exists in two forms, together with an additional fragment. The earliest manuscript is a draft with alternatives that was set down on the inside of an envelope addressed to ED and postmarked from Philadelphia. Franklin (1998, 1318) notes: "On a fragment of wrapping paper there is a portion of what appears to have been an intermediate draft, perhaps of more of the poem than the two lines remaining after the top and bottom had been removed (A515)." A fair copy of the poem is extant on a leaf from a student account book:

> We talked with
> each other about
> each other
> Though neither of
> us spoke -
> We were listening
> to the Seconds
> Races
> And the hoofs of
> the Clock -
> Pausing in front
> of our Palsied
> Faces

Time compassion
took -
Arks of Reprieve
he offered to us -
Ararats -
we took -

A516, F1506C/J1473/M621

The poem opens with the apparently paradoxical claim of talking without speaking. While Dickinson is often hard to understand, we nevertheless feel a sense of intimacy, a feeling that "she belongs to me alone" (see Chapter 8). The length of a particular discourse simulates the psychological distance between participants: the more distant the participants, the longer the discourse; the more intimate the participants, the shorter the discourse (Haiman 1985). The feeling of intimacy arises from shortened discourse, indeed to the point of not needing speech at all when two people are in perfect accord with each other's thoughts and feelings. I note that though the poem may appear to be "about" communication, in fact it opens with another unexpected word choice: not "we talked *to* each other" but "we talked *with* each other," an indication of shared experience. Neither, I suggest, need to speak because of their intimacy, but neither do speak because they are aware of the too-rapid passing of time. So instead of silence arising from intimacy, in this poem it arises from the rapidity of time passing. Both participants are recognizing the aging process in the other as time passes, recorded in the metaphor of the clock's ticking of seconds as a horse race.

That racing of time is reflected iconographically in the poem.[7] Phonetically, we hear the tick-tock of a clock repeated through the double occurrences of [k] in the six monosyllabic words: "talked–spoke"; "Clock–took"; "Arks–took." Their placement throughout the poem is also iconographic. Note that there is only one "tick-tock" in *talked–spoke* before the "Seconds / Races" in lines 7–8, a tick-tock that is lengthened by the distance between the words across the first five lines, whereas there are two after the racing of the seconds in *clock–took* and *arks–took*, a speeding up by number, and a speeding up by the placement of the third "tick-tock" coming immediately after the second. The length between the words *clock* and *took* in the second tick-tock captures time pausing between the otherwise quick alternation of the passing seconds.

The race against time is the tension of the participants in this poem who are constrained by a time limit on their meeting, a tension that translates as passion for this intimate and aging couple. Dickinson chose the variant "Time compassion / took" that reinforces the tick-tock phonetics for "Time's decision / shook" from the earlier manuscript. I now notice the two additional [k] words in "Seconds Races" (lines 7–8) and "Time compassion" (line 14), so that seconds become the tick of compassion's tock, and the placement of the tick-tocks is even further speeded up. Are the seconds

---

[7] Iconography is defined as an intentional (not arbitrary) relation between the elements of the art medium and the images or ideas expressed through them (M. Freeman 2020).

racing linked to time taking compassion, by pausing between the tock and the tick because of the participants' palsied faces?

Palsy, the *OED* informs me, is a disease of the nervous system that creates a certain paralysis, and "palsied" thus ranges in usage over being paralyzed, and thus is used figuratively to mean "deprived of muscular energy or power of action; rendered impotent." The word *pause* similarly refers to action stopping. The play between pausing and palsied resonates with the idea of a momentary lack of motion and conjures up the human construction of time's space between tick and tock. A pendulum clock does not in fact go "tick-tock," but has a regular succession of the same sound. It is human minding that divides the even succession into alternating two's. Frank Kermode (1967, 45) discusses the space between the tock and the tick, a temporal space too easily overlooked in the repetitive mantra of "tick-tock":

> The clock's "tick-tock" I take to be a model of what we call a plot, an organisation which humanises time by giving it a form; and the interval between "tock" and "tick" represents purely successive, disorganised time of the sort we need to humanise.

Such cognitive humanizing occurs as the couple hear the pausing of time between tock and tick as taking compassion on them in their own race against time.

Handwriting analysis places the poem at some time after the death of Judge Otis P. Lord's wife on December 10 (Dickinson's birthday) in 1877, and the possible onset of a romantic relationship between Lord and Dickinson.[8] Deaths are coming thick and fast at this time: her own father in 1874, Thomas Wentworth Higginson's wife in September of 1877, Samuel Bowles dying the next month after Mrs. Lord in 1878, followed by George Eliot in 1880, President Garfield and Dr. Holland in 1881, Charles Wadsworth, Emerson, and Dickinson's own mother in 1882, little Gilbert and Judge Lord himself in 1883. No wonder passion is intensified in this race against time.

The last tick-tock—*arks / took*—reveals the passion. The reference to "Arks" and "Ararats" is again a biblical reference, this time to the story in Genesis of Noah's ark surviving the flood. The covenant God made with Noah after the flood came to stand for the arks of the covenant in Hebrew biblical tradition. Offered as a reprieve by time, this couple seized the opportunity, whether of salvation from time or "taking" advantage of what time is theirs, in "taking" Ararat, where the ark finally lands as earth's permanence is restored, just as the poem itself preserves the couple's story in time. In this poem, the word *take* takes on even further possible meanings.

My analyses of the foregoing poems have focused on the life of words through their etymology and the ways in which they work to make Dickinson's poetry breathe, to animate it with the lifeblood of creativity. What I have been doing is, finally, the subject of another poem that again contains a biblical reference, this time to the

---

[8] Franklin mentions that the envelope on which the earlier version was written is postmarked from Philadelphia, though he doesn't say if the postmark carries a date.

disciple Thomas who wanted to see for himself before he would believe that Christ was resurrected from the dead (John 20, 24–29):

24 But Thomas, one of the twelve, called Didymus, was not with them when Jesus came.

25 The other disciples therefore said unto him, We have seen the Lord. But he said unto them, Except I shall see in his hands the print of the nails, and put my finger into the print of the nails, and thrust my hand into his side, I will not believe. [. . .]

29 Jesus saith unto him, Thomas, because thou hast seen me, thou hast believed: blessed are they that have not seen, and yet have believed.

The poem refers directly to "Sceptic Thomas":

> Split the Lark - and
> you'll find the Music -
> Bulb after Bulb, in
> Silver rolled -
> Scantily dealt to the
> Summer Morning
> Saved for your Ear, when
> Lutes be old -
>
> Loose the Flood -
> you shall find it patent -
> Gush after Gush,
> reserved for you -
> Scarlet Experiment!
> Sceptic Thomas!
> Now, do you doubt
> that your Bird was
> true?

A87-3/4, F905/J861/M427

The skylark is known for its liquid, warbling song. The silvery liquidity of the bird's song during flight has been a frequent theme of poetry. In Percy Bysshe Shelley's "To a Skylark," images of liquidity are paramount; the bird's tune is described in terms of "Drops so bright to see / As from thy presence showers a rain of melody." The notes that "flow in such a crystal stream" inspire the poet to create similar harmonies that "From my lips would flow." Shakespeare, too, identifies poetic creativity with the lark's song in sonnet 29: "Haply I think on thee, and then my state / Like to the lark at break of day arising / From sullen earth, sings hymns at heaven's gate." The music of Dickinson's lark spills out uncontrollably, "in / Silver rolled," metaphorically reflected in Dickinson's chemical knowledge of the silver globules (bulbs) of mercury that run uncontrollably all over the place and are hard, even impossible, to pin down. Mercury, also known as quicksilver, was named after the Roman god Mercury, the "silver-heeled," known for his speed and mobility. He is believed to have created the lyre and was the god of oratory and eloquence, language and writing. The lyre itself became a symbol of lyric poetry. Dickinson's poem is a paean to her belief in the power of poetry to continue to speak to us "when / Lutes be old -."

Often identified as an anti-scientific analysis poem, I note that until more modern technological procedures were invented for scientific exploration, the only way scientists could record and document the existence and nature of the world's flora and fauna was through death and taxidermy.[9] As in previous examples of word choice in Dickinson's poems, I expect "potent" in line 10 but get "patent": a lying open to view, a spreading wide, and also the idea of a discovery patented. "Split the Lark" becomes "Loose the Flood" as the image of music rolling out becomes the flowing of the life-force of blood. The ending of the poem is self-evident, with the explicit reference to the Doubting Thomas of St. John's Gospel. Thomas's need to feel for himself Christ's wounds in order for him to know that Christ is alive reflects the relation between the words made flesh and the spirit that keeps them alive, that makes them "true." Dickinson ends her poem with just the one word *true* in its final line, an emphasis that reinforces the power of poetry to speak truth.

The indivisibility of body and spirit, word and its vitality, are reflected in a late Dickinson letter-poem to Charles Clark (A726, F1627B/J1576/M730). It begins (lines 1–3):

> The Spirit lasts \
> but in what
> mode -

This question sets the stage for the letter-poem's subsequent contemplation on the relationship between the body and the soul, between the abstract and the concrete. The conceptual integration of our physical experience with our mental and spiritual consciousness is articulated and made known through language. Talking emanates from the abstract "Spirit" but can only occur through the concrete "Body" (lines 4–9):

> Below / the Body
> Speaks /
> But as the
> Spirit furnishes -
> Apart / it never
> talks -

---

[9] It was only in 1968 that two Russian scientists invented the method for perceiving the inside of a living cell in the electron microscope without killing it by fixing it in formaldehyde. Since then, the advent of computer simulations renders vivisection almost—and hopefully completely—obsolete.

The Western philosophical tradition, since Plato, has thought of the body as the "prison house" of the soul, a container from which the soul as an independent entity may be freed through death. For Dickinson, the relationship was much more intimate and mutually interrelated. As the letter-poem continues, the relationship between the abstract soul and the concrete body is explored further through an analogy with the music produced by a violin (lines 10–18):

> The Music in
> the Violin
> Does not emerge
> alone
> But Arm in Arm
> with Touch , yet
> Touch
> Alone - is not
> a Tune -

In these lines, Dickinson evokes the intimate connection between the instrument and the violinist, as the latter enfolds the violin within arms, chin, and bow to tactilely render forth the tune. The music does not exist independently of the touching of the violin, but the touching does not in itself produce the "Tune." Just as the body apart from its spirit cannot speak, the violin is mute without the spirit and touch of the violinist who calls the music forth. Neither can exist without the other, as the following lines, extending the analogy, confirm (lines 19–26):

> The Spirit lurks
> within the Flesh
> Like Tides within
> the Sea
> That make the
> Water live , estranged
> What would the
> Either be?

The sea wouldn't be the sea without its tides; the tides cannot exist independently of the sea. But what, exactly, are "tides"? They do not exist physically; just as the music is not a component of the instrument, tides are not a component of the sea, as salt is, or seaweed. They are a force that makes the water move, that makes the sea "live," just as the spirit is the life-force of the body. Dickinson's "disquiet" comes from the recognition that neither can exist without the other: "estranged / What would the / Either be?" In posing this question, Dickinson in effect raises the question that Western philosophy has never been able to answer—the question of the relationship between the concrete and the abstract. Unlike the concepts of earlier philosophers, philosophers like Plato, Aquinas, or Descartes who divided the abstract from the concrete, her three analogies—spirit/body; music/touch; tides/sea—suggest a different way of seeing, a way much more consistent with contemporary cognitive theory in its explanation of how concrete and abstract ideas are conceptually integrated.

Belief is not knowledge. And Dickinson had to know. In the second part of her letter-poem, she rephrases the question she asked in the first part (lines 27–35):

> Does that know -
> now - or does
> it cease -
> That which to
> this is done ,
> Resuming at a
> mutual date
> With every future
> One?

Here, Dickinson seems to be suggesting that there are only two alternatives: that, estranged, both cease, or that they resume their future mutuality. Why does she not consider the third possibility, the one that her own tradition and Christian culture presented: the continuance of the soul beyond the death of the body? The "Reply" she gives at the end of the poem is no reply at all to the metaphysical question of life after death, but it does answer this last question (lines 36–52):

> Instinct pursues
> the Adamant ,
> Exacting this
> Reply ,
> Adversity if it
> may be , or
> Wild Prosperity ,
> The Rumor's
> Gate was shut
> so tight
> Before my Mind
> was sown ,
> Not even a
> Prognostic's Push
> Could make
> a Dent
> thereon -

The unverified information—the "Rumor"—of immortality cannot be predicted or foreknown—prognosticated. The world she saw elsewhere as "Paradise," that enclosed garden of delight, is a container that cannot be breached by the mind that is contained—"sown" as the seed—within it. Her refusal to accept with confidence the easy faith of her father's in the soul's immortality, however much she wanted to, lay in her knowledge of human cognition, that knowledge of the abstract cannot exist apart from the concrete, that understanding is, ultimately, embodied.

Words made flesh are infused with the animating life-force of the spirit throughout Dickinson's poetry. We do not have Higginson's first letter to Dickinson in reply to her question if her verse were alive, if it breathed. One would hope, as suggested in Dickinson's second letter to him, in which she noted that he asked for other poems, that he assured her that it did, in spite of "the surgery" (L261). In her first letter, she asked Higginson to tell her "what is true." Her own poetry reveals she had no need to ask: her words made flesh are indeed "full of grace and truth."

Poetry is thus made up of words: words playing off each other, words working with and against each other, words refracting their many potential meanings. From a cognitive linguistic perspective, the traditional notions of "denotative" and "connotative" meanings no longer make sense. Meanings do not reside in the word, either denotatively or connotatively. They are encyclopedic, created from the dynamic interactions that take place among writer, text, and reader. That is why a poem can mean so many things to so many people. Words have networks of potential meanings, so that a word in a poem reverberates with and against its other possible elaborations and extensions, depending on the context of the utterance in which the word is placed (motivated by the intensions of the writer), and the experiential background the reader brings to the text.

Bringing a Poem to Life

Long before I encountered linguistics or cognitive theory, I experienced the beginnings of what was to come: an abiding interest—the question of how one encounters literature, and specifically, how to read poetry—in two transformative episodes. The first occurred in grammar school as a teenager in England. We were given William Wordsworth's poem, *The Prelude*, to read. I didn't much like it. Except for the memorable incident where the boy was affected by "grave and serious thoughts" from a moonlight escapade on a lake, I found it superficial and boring. It was not until I encountered the poem once more at university that I realized the poem I had read had been abridged: all the philosophical passages had been removed. The poem for me then finally lived up to its subtitle: "Growth of a Poet's Mind." The discovery was earth-shattering. That taught me two things: (1) never to read a work of literature, whether prose or poetry, in abridged form, and (2) poetry involves far more than a simple reading.

The second episode occurred in my final year as an undergraduate in English and Philosophy at the University of Manchester. I had chosen as my special subject a course in American twentieth-century fiction. I knew nothing about the United States. In school, the closest we ever got to America was the story of Wolfe climbing the Heights of Quebec to defeat Montcalm—and that was Canada. In retrospect, the studied avoidance of any mention in my grammar-school years of the United States, in history, politics, or literature, was staggering.[1] My university instructor, Geoffrey Moore, was the first full-time lecturer in American Literature from 1955 to 1962 at a British university.[2] The first book he assigned us was William Faulkner's *The Sound and the Fury*. It was the first—and only—book I have read cover-to-cover in English without understanding a word. It made me realize two more things: (3) knowing a language does not automatically entail comprehension, and (4) I knew absolutely nothing about America. That final realization drove me to spend my postgraduate year at Smith College in their newly created American Studies Diploma Program.

---

[1] When I told the Professor of Philosophy that I was going to the United States for a postgraduate year of study, she said "Don't bother to read philosophy there—there isn't any," though she herself had studied with Alfred North Whitehead and taught for a while at Columbia University.

[2] Geoffrey Moore, while a member of the Faculty of English, was informally associated with Marcus Cunliffe's American Studies group, the University of Manchester being the sole institute of higher learning in England at the time to feature study of the United States. Moore's (1964) book was the first anthology of American Literature to appear in England.

There I encountered my first experience of Emily Dickinson's poetry, in a seminar conducted by Ellen Smith. That confirmed for me the fact that though I had read poetry all my life, I lacked the linguistic skills needed to understand how it worked.

Over the years, through studying linguistics and then, later, research in cognition, I have changed the way I read poetry, from interpretation to experience. As a result, this chapter addresses the question of how one can experience a poem through understanding what it means to read a poem cognitively. By that, I don't mean simply conceptual interpretation. I mean how we as embodied animate organisms experience what a poem is doing. The notion of embodied cognition implies the phenomenology of our being part of the world; that is, the way our entire bodily organism is animate as it both affects and is affected by the "affordances" of its environment. Embodied cognition is thus also embedded, extended, and enactive, what researchers call 4e cognition (Newen et al. 2018). When we encounter a poem, we engage in active sense-making through all our cognitive functions: sensory, motor, emotive, and conceptual.

A poem is not something to pick up and absorb in one reading. To read beyond the superficial level, you must enter a poem and allow the poem to enter you. In other words, reading a poem cognitively is to live with it, and let it live with you. Just as Dickinson breathed life into her poems, we as readers respond to the poems through our embodied perceptions and bring them to life for ourselves. Only then can we be assured that whatever interpretations we make are possible within a poem's total gestalt.

As living, animate organisms, our own subjectivity is rooted in the integration of the sensory-motor-emotive processes that underlie our conscious, conceptual awareness as we actively interact with the world outside ourselves. The affectiveness of a poem comes primarily from what we hear: the sensory relation of sound to sense (MacLeish 1960). When spoken aloud, a poem's inflections, intonations, pauses, and so on, animate the sensory, motor, and emotive experiences that lead to conceptual awareness and underlie the language of its utterance.

The following readings of two Dickinson poems conceptually summarize the active process of feeling the emergence of a poem's being. I hope they will suggest strategies for engaging with a poem rather than simply reading for meaning.

1 Cognitive Construal

In some cases, a Dickinson poem can be so linguistically complex that cognitive analysis must precede a linguistic one even to begin to make sense of what is going on. Take, for instance, the poem: *To undertake is to achieve* (A88-3/4/A493, F991*AB*/J1070/M455). Published versions of the poem give the first line as "To undertake is to achieve," giving the impression that it is a complete sentence in itself. The line break after "achieve" reinforces the idea of closure, with no indication of what or how something is being undertaken. Searching the internet for the line, *To undertake is to achieve*, produces many citations used for inspirational purposes. Is Dickinson adopting the modern pedagogical idea of giving "an 'A' for effort"? When the entire poem is given online, instead of an inspirational response, you get the following from someone called Shannon: "didnt like the poem one bit. . . i dont get it really. . . its a little stupid if you ask me."[3] What is going on?

The poem exists in three manuscript versions. With the exception of the variant word choice "Sources" for "Natures" in line 12 that appears in two of the versions, there is no indication that Dickinson was dissatisfied with the poem's structure. It is, I think, notable, that in all three versions, the words "undertake" and "achieve" are separated by a line break. This in itself suggests resistance to closure, which, as we will see, is important to understanding the poem as a whole. Since manuscript comparison is not my objective here, I shall use the version that appears in fair copy in set 7, noting and including the variant word choice from the other versions:

> To undertake is to
> achieve
> Be Undertaking blent
> With fortitude of
> obstacle
> And toward Encouragement
>
> That fine Suspicion
> Natures must
> Permitted to revere
> Departed Standards
> and the few
> Criterion Natures [*Sources*] - here -
>
> A88-3/4, F991A/J1070/M455

This poem is one of the most difficult in her entire collection. Reading it aloud reveals particular problems with the last stanza. Its elliptical nature is unsurmountable by syntactic analysis, with seemingly an infinite array of possibilities. I will just mention the semantic ambiguity of the word "Suspicion," read differently in the only two literary interpretations that I've been able to find. Ruth Miller (1968) sees the poem, one copy of which was sent to Thomas Wentworth Higginson, as "an artful disguise of her attitude toward her mentor": that to undertake writing poetry such as hers is already an achievement. Miller reads the word "Suspicion" as meaning mistrust and claims that the stanza refers to "critics ['Natures'] who of course must be suspicious if all they permit themselves to admire [revere] are old poets who established the traditional forms ['Departed Standards'] or those few successful contemporary poets ['Criterion Sources'] who are assumed to be the criterion of good poetry" (7). Richard Sewall (1948), noting that the poem was identified as an "experimental" or "imperfect

---

[3]   The comment was made by Shannon from the United States on November 16, 2004 to the printing of the poem on the American Poems website http://www.americanpoems.com/poets/emilydickinson /11022. Accessed July 6, 2022.

START ———————————→ PATH ———————————→ END

SOURCE   GOAL

undertake   fortitude   encouragement   achieve

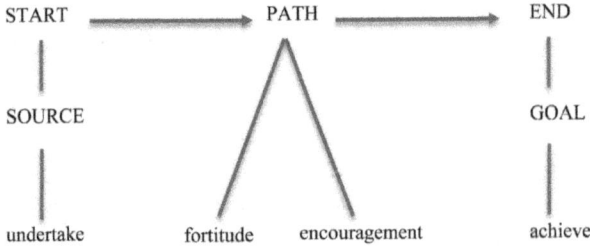

Figure 7.1 PATH schema.

work" when first published in the *New England Quarterly* in 1932, identifies its theme as being a notion beloved of the nineteenth century: "nobility or striving, even if achievement is denied." Sewall reads "Suspicion" as meaning hint, clue, or belief, and suggests a syntactic reading in which the first lines of each stanza, "To undertake is to achieve" and "That fine Suspicion / Natures must" (Sewall adds the word *cherish* to complete the thought), make a complete statement which is qualified or developed by the rest of the stanza. Sewall concludes that the poem's fault lies in its too-tight rather than too-loose organization. I am not satisfied that either Miller or Sewall has fully listened to the voice of the poem. Instead, I suggest that a cognitively focused analysis can begin to show what the poem might be doing.

First, I note the continuation of the sentence in the first stanza that provides the conditions for an undertaking's achieving success. Its linear movement has PATH as the image schema underlying its structure (Figure 7.1).

The focus is on the hypothetical conditional, represented by the path itself: only if one shows fortitude in the face of obstacles and is emboldened by encouragement will one achieve what one undertakes. The first stanza is fairly straightforward in this respect. But what it means is that attention is directed not to the start or end of the JOURNEY metaphor, but on the path, the doing, not the result, being in focus.

Although Dickinson frequently employs the PATH schema, her poetics are determined by a non-linear schema (see Chapter 13). That is, the figure of a Dickinson poem, unlike that of a Frost poem that "begins in delight and ends in wisdom," is one of elaboration and expansion through change or transformation.[4] Thus, in the second stanza, Dickinson focuses on the conditional elaboration of lines 3–6 by expanding on the nature of what is needed for achievement to take place. She is not simply saying that there is an immediate cause-effect relationship between undertaking and achievement. Her poem expresses the cognitive semantic notion of *force dynamics* in any event structure complex that involves causation. In cognitive semantics, force dynamics refers to the range of relations that one entity can bear to another. Entities in themselves have intrinsic force, and these forces can have opposing tendencies, with one entity resisting the opposition or blockage, and the other overcoming that

---

[4]   Robert Frost's quote comes from the succinct and wonderful statement of his poetics in his essay, "The Figure a Poem Makes," which he added as an introduction to the second edition of his *Collected Poems* in 1939.

resistance. This range, in Leonard Talmy's (2000, 10) words, "further includes the presence, absence, imposition, or removal of blockage to one entity's intrinsic force tendency by a second entity. In force dynamics, causation now appears within a larger conceptual framework in systematic relationship to such other concepts as permitting and preventing, helping and hindering."

This movement can be seen in *To undertake is to achieve* in the semantic networks of Dickinson's linguistic terminology. As a poet, she is making her words work by reflecting off each other with their own kind of force dynamics, and as a result, a voice is emerging. The terms Dickinson uses indicate the inner resources needed to engage in any undertaking. They reflect the hazardous nature of venturing forth without the assurance of success.[5] In the first stanza, the word *undertake* itself means to venture or hazard, as indicated in her Webster dictionary, and *obstacle* highlights the dangers of such hazard. The terms *fortitude* and *encouragement* both introduce the notion of the force dynamics that are the basis or source of courage. Etymologically, *encourage* comes from the French to take heart, and carries the idea of potential emboldenment. *Fortitude*, as Webster notes,

> is the basis or *source* of genuine courage or intrepidity in danger, of patience in suffering, of forbearance under injuries, and of magnanimity in all conditions of life. We sometimes confound the effect with the cause, and use fortitude as synonymous with courage or patience; but courage is an active virtue or vice, and patience is the effect of fortitude (my emphasis).

The variant "Sources" for "Natures" in the second stanza makes clear this emphasis on the force dynamics needed for any achievement to occur and indicates that the word *nature* is used in its sense of the essential quality or attribute of a thing or person. Lines 3–6 of the first stanza thus create a hypothetical blended space in which encouragement is the emergent structure of blending undertaking with fortitude of obstacle.

This encouragement can only be achieved, however, if our essential nature is "permitted" to undertake with fortitude "that fine Suspicion." As noted earlier, the word *suspicion* is semantically ambiguous. However, within the context of Dickinson's poem, the phrase "fine Suspicion" maps onto the terminology of the first stanza in meaning the discrimination ("fine") of both imagining the possibility of something as yet unproven (Sewall's "hint") and the fear that attends such uncertainty (Miller's "suspicious") that are the necessary essence of any undertaking of the unknown. We are now beginning to hear the poem's voice.

The syntax of the second stanza is, however, still problematic. Should we read the first stanza as a complete sentence, with "That fine Suspicion" somehow in apposition to it, as Sewall's analysis suggests? Or should we read "That fine Suspicion" as the object of "achieve," coming after the conditional force dynamics expressed in lines 3–6? Or

---

[5] Many poems illuminate this theme. See, for example, *Finite to fail, but infinite to venture* (A86-9/10, F952/J847/M443), *We never know how high we are* (A512, F1197/J1176/M555), *Lest this be heaven indeed* (A88-9/10, F1000/J1043/M458).

should we see at least part if not the whole of the second stanza as a continuation of the parenthetical hypothesis of lines 3–6? As readers, we want to parse that final stanza. But Dickinson won't let us do it. She prevents it by the extreme ellipses of any and all grammatical connections. Why does she do this? Imagined outcomes are too easy. Dickinson is suggesting just the opposite of the inspirational: that achievement is not easy, not just a question of undertaking. But the undertaking itself will determine the outcome, depending on the inner resources we bring to the task.

We don't know whether we can accomplish our objectives until we try. But how do we determine the best route to take? In Dickinson's terminology, what standards or criteria do we adopt? I thought about this poem when I was part of a group clearing a hiking trail after the ravages of winter. We came across a large tree blown down at an awkward angle across the path. With only a bow saw to hand, the task of removing the tree seemed impossible. Hikers could manage to clamber over it (and did), though it would prove a serious impediment to cross-country runners in training at our local school. I quoted aloud the first stanza of Dickinson's poem. We were determined to try. Having worked out the force dynamics in the probability that the saw would bind and the danger of vector tension in the log giving way unexpectedly, we set out from that departure point and began to work on removing the fallen tree. As we worked, I began to realize the effect of Dickinson's second stanza. To undertake means to take a risk, to depart, to set out from a known position to an unknown future, with the requirement that we commit our essential natures to the task.

Dickinson's poem, in projecting back to the starting point or source rather than forward to the ending point or goal, is doing what it is saying. It is creating a complex blend in which the poem's language iconically creates the obstacles and encouragements experienced through respecting our essential natures that are the source of anything we undertake. Accomplishment isn't the point. Attempting is. What I now hear in the second stanza is the voice that is plunging in to undertake a journey into the interior self, a voice that struggles with the force dynamics of closings and openings, obstacles and permitting, that such an undertaking involves. As we struggle with the complexities of the second stanza, we are submerged in the very undertaking toward understanding that Dickinson is getting at. The online respondent who didn't like the poem because she didn't get it and therefore thought it stupid is not engaging, not listening to the voice of the poem, not taking the labyrinthine journey the poem invites us on (Leão 2002).

2 Cognitive Analysis

Roman Ingarden (1973, 368) identifies the following order of procedure in a cognitive reading:

1. the aesthetic experience of an ordinary reading, which raises questions concerning the peculiarities and execution of the work;

2. pre-aesthetic reflective cognition, which forces an attentive and cautious reading, leading through more concrete and adequate examination to a synthetic appreciation of the work which gives new insights, that

3. enable aesthetic concretization, in which "places of indeterminacy" can be filled out to express the effectiveness of the aesthetic experience, which in turn

4. "must belong to that domain of values which lies within the scope of the possible realizations of that work of art" and not just those related to the reader's experience.

Like *To undertake, To make routine* (H05, F1238/J1196/M511) is also puzzling and therefore not surprising that Mabel Loomis Todd marked it as "No" when she was considering poems to publish in the early editions. It is not easily understood on a first or even subsequent reading (Figure 7.2).

The following steps outline my several engagements with Dickinson's poem, not necessarily in the order in which I list them, but cumulatively and sometimes simultaneously as I began to live with the poem and let it speak to me.

Step 1: Intuiting Aesthetic Emotion on a First Reading

First readings provide an immediate response of engagement or otherwise on the part of the reader. Like many of Dickinson's short poems, this one is not transparent on a first reading. The first eight lines seem straightforward enough, but the final eight lines puzzle. However, I intuitively feel a sense of consolation and reassurance, even in the face of language that suggests otherwise. The rhymes of *cease* and *grace* serve to link the idea of ending as something desired, while the rhyming of *repair* and *more fair* that brings the poem to its ending strikes the more positive note that may have contributed to my initial feelings of consolation and reassurance.

In the first four lines, I note that the speaker apostrophizes an addressee with the admonition "Remember." The opening phrase, to make, can be understood either as "if you want to make" or "in order to make." Is routine something to be appreciated or deprecated? Why should the thought that routine can stop turn it into a stimulus? A stimulus for what? To make routine something other than it is? Or to make routine itself stimulating? Is *capacity to terminate* an example of the middle voice, in the sense that routine itself is capable of ceasing, or does the phrase refer to the fact that we can end routine whenever we want? Why grace?

If these eight lines don't present problems enough, the next eight are worse. Does the phrase "Of Retrospect" belong to "Grace" or does it start a new thought? What is/are the subject(s) of the verbs *departed* and *become*? Is *departed* a past participle or the simple past tense of a main verb? Is *that* in line 11 a demonstrative? a relative pronoun? or does it introduce a complement clause? Why should something "more fair" be regretted in that parenthetical "Alas"? How did an arrow get into this poem?

These difficulties may cause casual readers to throw up their hands and say "Why bother?" But if the poem intrigues, the challenge is to let it emerge and reveal its "being."

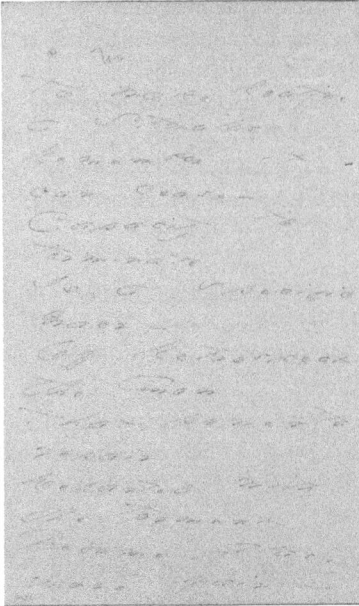

To make Routine
a Stimulus
Remember it
can cease -
Capacity to
terminate
Is a specific
Grace -
Of Retrospect
the Arrow
that power to
repair
Departed with
the torment
Become , Alas ,
more fair -

Figure 7.2 *To make routine* (H205, F1238/J1196/M511) ms_am_1118_3_205_0004 Courtesy Houghton Library, Harvard University.

Step 2: Looking Closely at the Language of the Text

I create several copies of the poem to use as worksheets in order to identify the various features, correspondences, and repetitions I find. A linguistic analysis includes three categories: structure (syntax), sense (semantics), and sound (phonetics). So first I look at the poem's structure. On a macro level, the sixteen-line manuscript poem divides into two parts, both comprised of eight lines. Immediately, I am led to consider the second part an elaboration or commentary on the first part, Dickinson following the characteristic format of a biblical passage (Berlin 1985) rather than a linear trajectory of idea development.

I can see that the difficulties of the last eight lines lie primarily in their syntax. Often, different readings of a Dickinson text result from resolving ambiguous syntax in only one way. But recognizing the possible ambiguities can also allow the reader to hold more than one reading at the same time. I could, like my Tübingen linguistic colleagues do, run through all the syntactic possibilities one could give to these last eight lines, including attaching them syntactically to the end of the first part.[6] Such

---

[6]   The University of Tübingen A2 project, "Interpretability in Context," is part of the larger research project SFB 833: Bedeutungskonstitution—Dynamik und Adaptivität sprachlicher Struktur (http://www.sfb833.uni-tuebingen.de/). Its recent publication is Matthias Bauer et al. (2020) on Dickinson's grammar.

linguistic analysis explains why readers come up with different meanings. But a linguistic analysis in itself cannot determine which formulations cohere with other aspects of the poem. Instead, I turn to notice other language strategies before trying to make sense of the syntax.

Dickinson's love of philology made her particularly sensitive to the possible meanings of words arising from their etymology. Although the breadth and depth of her reading was immense, which would have given her ample resources for suggestive meanings of words, Cynthia Hallen's online *Emily Dickinson Lexicon* (edl.byu.edu) that includes Dickinson's Webster's 1844 dictionary is a place to start, though I also consult the *Oxford English Dictionary* (*OED*) with its comprehensive definitions and examples of usage through the centuries. So, even though we mostly already "know" the meanings of words as we read, knowing what Dickinson's lexicon and reading afforded in determining specific contexts for meaning enriches my understanding. I therefore start by creating a worksheet of definitions from Webster 1844 for all the significant words in the poem.

Step 3: Identifying Prosodic Features

Metalinguistically, the emotional weight of a poem lies in its prosody, its sound patternings, rhythms, phrasal groupings, line breaks, repetitions, markings, and so on. Overall, there is a certain symmetry across the two parts of the poem. I note the three infinitives: *to make, to terminate, to repair*; the appearance of *terminate* and *repair* on lines of their own, making the word *to* appear twice at line end; the parallelism of *capacity to* and *power to, can cease* and *terminate, remember* and *retrospect*. I note that the noun phrases *a stimulus, the arrow, the torment* also appear on separate lines, as do the end rhymes *cease/Grace* and *repair/more fair*. As a result, these phrases are foregrounded, demanding special attention. Endings and aftermaths seem to dominate as themes.

Unlike many of Dickinson's poems, this one has only five markings: two horizontal, fairly long lines after *cease* and *grace*; and the remaining three in the final two lines, two rising slides after *become* and *alas*, and the expected downward, longer marking at the end (see Chapter 3). Another striking visual feature are the long, downward-sloping slants forming the letter *t*. Dickinson's usual practice throughout her poetic corpus triangulates this letter to mark a capital by joining the crossbar to the stem, and in lowercase words often adds it after its stem. But here, in both capitals and lowercase words, the long, downward-sloping bars across the stem of the letter *t* stand out. Like the falling slide, the authoritative downward slants call emphatic attention to the context in which they appear. It is no accident that among the other predominating consonants, the phonetic [t] sound resonates. It occurs four times in the opening three lines to disappear in the fourth "can cease -" and then again four times in the next two lines to disappear in the following two: "Is a specific / Grace -."

The following eight lines have a similar, though variant, pattern: two occurrences of [t] in line 9 followed by none in the next line; three in line 11 followed by none in the next line; then three in the next two lines, followed by none in the last two lines. So you have a pattern of presence followed by absence of phonetic [t] in both parts of the poem, with the same number of eight in both parts, but with increasing disappearance over both parts.

Consonantal sound patterns predominating are [r], [t], [k], [s], [m], and [p]. Note that [r] occurs four times at word onset (*routine, remember, retrospect, repair*), six times internally (*terminate, grace, retrospect, arrow, departed, torment*), and five times at word end (*remember, power, repair, more fair*). The effect is to link sound and sense suggestively. Without presenting all the details of my analysis of the other sound patterns, here is a list of my findings that reveal how sound consonant patterning supports the poem's overall symmetry.

- [p] and [m] never occur in the same line.
- [m] drops out after *terminate* in line 6 and doesn't reappear until *torment* in line 14 (note the same pattern of [t]-[m]-[t], with no [k] or [s] or [p] in the paired lines).
- The only lines in which only one of the sounds [t, k, s, m, p] occur are line 8, *grace*, line 12, *repair*, and line 16, *more fair*. These do not include [t] or [k].
- [p] occurs with [s] within a word twice in the first part and only once in the last, with two occurrences without [s].
- [m] occurs only in the first three lines and the last three lines, except for line 6 on *terminate*.
- [k] and [s] predominate in the first part; they disappear altogether in the second part until line 15 when their order [k], [s] chiasmically mirrors the [s], [k] of line 9, linking *retroSpeKt* with *beKome, alaS*. Note also the disappearance of [t] from the second of these two lines.
- *the arrow* in line 10 is sore-thumbed: it sticks out by being the only line in the entire poem that contains none of the sounds [t, k, s, m, p]. Is this the "eye" of the poem?[7]

When rearranged as in the print editions into eight lines, the poem reveals the iambic common hymn meter of 8-6-8-6, with one significant exception. Just like the reduction and disappearance of sounds between the two parts, the second four lines lack a final stress position to give a pattern of 7-6-7-6. The line breaks reinforce the foregrounding of the shorter lines, providing a parallel between *the arrow* and *the torment*. That makes me look back at the first part to see that an equivalent parallelism occurs in *a stimulus* and *terminate*.

Step 4: Recognizing Cognitive Import

Already the previous three steps have been urging toward a cognitive reading of the poem. First, then, I consider the possible import of sound and sense before tackling the most difficult syntactical structure of the last eight lines.

---

[7] One of Haj Ross's (2000) most striking examples of a poem's "eye" is the line from Blake's *Tyger, Tyger, burning bright*: "Did He who made the lamb make thee?" which becomes the key to the poem.

We have already seen in step 3 how sounds appear and disappear, how they interplay with each other as the poem proceeds, how they link lexical items with each other. Images of departure and termination, of memory and retrospect, of pain and reparation are reinforced by such sound patterning. To cite just one example, note how the sound [r] appearing at word onset in *routine* in the first line occurs in word middle in the sore-thumbed line 10 (*arrow*), beginning and end in *repair*, and then repeats on the last two lines of the poem at word end (*more fair*), associating the four phrases in progressive movement.

Senses of the chosen words in the order of their repetitions enrich their meanings. Thus, *cease*, to stop, is followed by *terminate*, to end, to culminate in *departed*, which in Webster's definition of the past participle is simply "gone from, vanished, dead," again a progressive movement, this time from the possibility of starting again to its denial in the termination of death. Similarly, the passive possibility of *capacity*, associated with *terminate*, in the first part becomes the active *power*, associated with *repair*, in the last. Although this seems to imply positive restitution, it is repudiated by the departure of power "with the torment."

That leaves us with the puzzle of syntax. My first reading suggested that the final eight lines serve as a commentary and development of the first eight. However, a preliminary reading of these lines, though syntactically possible, does not illuminate the questions I raised at the beginning: Does *of retrospect* start a new sentence, or is it a continuation of *a specific grace*? What is the grammatical form of *that* in line 11? What does *torment* refer to? What is the subject of *become*? Why *alas* in the final line? And, finally and ultimately, what is an *arrow* doing in this poem?

Thinking of what I have learned from the poem so far, and remembering Blake's admonition, I look more carefully. Had lines 7–8 read "*the* specific / Grace -," the determiner would have pointed to what distinguishes the kind of grace being mentioned; that is, the grace of retrospect. But it doesn't. So I conclude that the fairly long horizontal mark after "Grace" separates the two parts, and that the three horizontal markings all signify the end of sentences. That leaves the last eight lines as one long sentence, again, a reduction of sentences from two in the first part to one in the last. The greatest puzzle for me as I learned to live with the poem and let the poem live in me was to understand what the subject of the verb *become* is in terms of the total import of this last sentence in relation to the whole. I discover that it all hinges on the syntax of *that* and the presence of *the arrow*.

Without exhausting your patience in following my many steps and retreats, here's what I finally came to realize. The tone of the first part is upbeat. Just remembering the ability to cease from routine can raise one's spirits (one of Webster's definitions of *stimulus*). Such a memory is thus *a specific grace*. It's a statement of anticipating possibility that breathes life into what is otherwise dull and deadly routine. In contrast, the tone of the last part is downbeat. Rather than being a commentary and elaboration of the first, it is a counterfactual, considering the consequences of ceasing from routine. One is invited to look backward, to see that, should one actually terminate routine, it may not be restorable.

Dickinson often creates focus by pre-posing items, so that the normal order of lines 9–10 would be *the arrow of retrospect*. By putting "Of Retrospect" first, the *specific grace* resulting from the possibility of ceasing from *routine* becomes the *arrow* of *retrospect*. In other words, looking back makes one realize that what seemed to be a gracious gift was actually an arrow that causes the end of routine's torment, making the *power to repair* impossible. This *cause-effect* relationship characterizes the underlying schematic force of the poem.[8]

Although *that* in line 11 can be read as a demonstrative or relative pronoun, it results in *the arrow* being the subject of *become*. This doesn't make sense. Instead, I hear it as the introduction to a complement clause, indicating "such that," resulting in *power to repair* as the subject of the verb *become*. This explains the *alas*. A paraphrase would then read: "The arrow of retrospect is such that the power to repair that departed with the torment [that the arrow terminated: that is, routine] becomes, alas more fair." The irony is thus established: the promise of escape from routine can happen only by the instrument of ultimate termination, with no possibility of restoration.

In brief, my cognitive reading notes that the whole poem breaks into two parts: anticipation and aftermath, two critically documented aspects of Dickinson's poetics in the literature.[9] The anticipatory possibilities of relief from routine, the endless round (etymologically derived from *rota*, wheel) of regular habits and activity, expressed through the memory that it can always cease, is countermanded by recognition of the aftermath. Although engaging in routine may seem torturous, the fact that it is continuous means that it has the power of restoration. Once it is gone, in retrospect, *routine* seems "more fair" in comparison with the *stimulus* that can lead to unrecoverable loss.

I suggest that a cognitive reading is not simply another literary analysis. Rather, it provides the grounding for literary interpretation. For example, one could generalize the themes of this poem by elaborating the roles of memory or regret, or by linking it to other poems with similar semantic networks, such as routine and round. Another poem, *The soul should always stand ajar* A90-9/10, F1017/J1055/M/463), has the same format as the two discussed: an opening line followed by what would happen otherwise.

Step 5: Evaluating the Poem's Success

What a poem is saying is not what a poem is doing. A paraphrase cannot capture the feeling a poem creates when what the poet "sets out to do," in Wallace Stevens's words, succeeds. For me, in spite of my coming to understand the gist of the poem, I am still

---

[8] In his research on the history of the arrow in literature, Cristóbal Pagán Cánovas (2011, 564) formulates the cause-effect network of conceptual integration as follows:

> This generic network includes at least the following mental spaces: causation, a relevant human experience, causal tautology, this same relevant experience as a general cause, an agent causing the typical result of this experience, an agent causing a typical result of a different event (recruited as an analog for cultural reasons), and the blend, where the experience in general is shaped as an agent of the analogous event.

[9] The notion of possibility, of things not yet realized, permeates Dickinson's poetry. For the preference of anticipation over realization, see D. Freeman (2005). For the notion of aftermath, see D. Porter (1981: "The Crucial Experience," 9–24).

left with the way the poem points toward the feeling of lost possibility. Had the second part begun with "In retrospect," it would simply have produced a cautionary note, along the lines of "Be careful what you wish for." That this poem is doing much more is captured in that little word *of*. It is looking back on the consequences of actually ceasing routine that makes specific grace the arrow—the cause—of recognition of what has been lost. This poem works for me because it "realizes" its statements about routine and reparation, retrospect and regret through its prosodic structure.

A cognitive analysis enables me to perceive why I had the intuitive feelings of aesthetic affect of consolation and reassurance on first reading the poem. Abram (1996, 158–9) gives me a clue to the poem's cognitive effect in his description of Apache *'agodzaahi* ("that which has happened") stories, which always begin and end where the events in the story actually occurred ("It happened at . . . "):

> The telling of any such tale today is always prompted by a misdeed committed by someone in the community; the *'agodzaahi* story, precisely told, acts as a remedial response to that misdeed. Thus, when an Apache person offends the community by a certain action, one of his or her elders will wait for an appropriate moment— perhaps at a community gathering—and will then "shoot" the person by recounting an appropriate *'agodzaahi* story. Although the offender is not identified or named aloud, he or she will know, if the "arrow" (the tale) has been well chosen and well aimed, that he is the target; he will feel the story penetrate deep beneath his skin and sap his strength, making him feel ill and weak. Then the story will begin to work on him from within, making him want to change his ways, to "replace himself," to live right. And so his behavior will change. Yet the story will stay with him. For he will continually encounter the place in the land where it all happened.

I don't know if Dickinson would have learned of such a thing, or anything like it, but even the language of Abram's description resonates with her Routine poem (remediation, deep within, working from within, replacement, continuity, encounter). Read in the light of an *'agodzaahi* story, the poem makes perfect sense. The topography of Dickinson's poem is the mind. The reader who is trapped in the first stanza by thinking that stopping routine is a good thing is "shot" in the second by the unfolding realization that it isn't. The arrow (telling the story reminding one of the possibility that routine can end) has the capacity to terminate one's attitude about routine and the power to repair one's feeling of torment as we endure routine. Remembering that routine can cease transforms it into something stimulating, and thus can change our minding about it, causing the torment to depart. But if it does, our retrospective thoughts about routine will become "Alas, / more fair -" (thus explaining the alas), and we will regret the ending of routine by recollecting that it is a good thing which will thus restore (repair) our attitude toward it.

Given the causative elements signified by the infinitives in the possibility *to terminate* in contrast to the ability *to repair*, ironically and characteristically for Dickinson, anticipation always wins over realization. Cognitive study, however time-consuming and laborious, reaps its own rewards in understanding and appreciating a poem that "works."

8

Intimate Discourse

Emily Dickinson is unique among major poets for being both very obscure and very popular, not only among academics but in the world at large. This contemporary paradox mirrors a historical one. Despite the fact that a substantial scholarly apparatus has developed for the study of her poetry, both nineteenth- and twentieth-century critics have used an evocative lexicon of negation to describe their experience of reading a Dickinson poem: lack, gap, space, absence.[1] This quality of what is not there, its foregrounded presence of absence, is implicit in the critical responses to her poetry from both her very earliest reviewers and contemporary Dickinson scholars.

But although critics for more than a century have described their various responses to the seemingly paradoxical impression of obscurity and popularity, they have not explained it. Why does Dickinson's extremely difficult poetry nevertheless cause many people all over the world to experience feelings of attachment and identification? Dickinson is widely read because she creates a sense of participation in which she iconically represents a crucial fact about intimate conversation: the closer the relationship between participants in a discourse, the more implicit the speaker's discourse becomes and the less help the hearer needs to process it.

Precisely this contrary-to-fact intimacy between persona and reader accounts for the "difficulty" of a Dickinson poem, difficult because it leaves to silence what formal discourse among strangers normally would demand. This very same intimacy accounts for the closeness and strong sense of identification that Dickinson's readers feel, even when it is hard to determine what a poem might be saying. In intimate discourse, meaning resides as much in the nuances of what is not said as in what is. This fact presents a great challenge for readers of Dickinson's poetry.

Dickinson draws the reader into a feeling of intimacy by adopting several cognitive discourse strategies that iconically decrease the distance between writer and reader. These include colloquial expressions, shortened discourse, omitted reference, shared context, and shared knowledge.

---

[1] References are too numerous to cite in full. Nineteenth-century reviews are full of such terms (Buckingham 1970). The title of Inder Nath Kher's (1974) *The Landscape of Absence: Emily Dickinson's Poetry* is suggestive, as is Suzanne Juhasz's (1983) *The Undiscovered Continent: Emily Dickinson and the Space of the Mind*. Sharon Cameron (1992) has an extended discussion of this quality in Dickinson's poetry.

Consider, for example, the following two sentences that share the same meaning:

1. The statement I previously communicated to you has proven to be accurate.
2. I told you so!

The first sentence sounds pedantic and stuffy, the second informal and pithy. Notice that the first is three times as long as the second in the number of words and over five times in the number of syllables. They share only two words in common: *I* and *you*. The first sentence uses words derived from Latin (*statement, previous, communication, accuracy*). The word *tell* in the second sentence has only one syllable, pointing to its Germanic origin. The phrase *previously communicated* indicates that some discourse has been delivered at some point prior to the present moment. In its past tense form, *told* is all that is needed to indicate that a telling took place in the past. The word *so* not only refers to what was said but to its truth. It thus replaces the entire string, *the statement has proven to be accurate.*

From these examples, it would seem that short style is always better than long style. This is not necessarily the case. The first sentence spells out on the surface what is underlying in the second. Such spelling out can indicate subtle differences in meaning. For example, the preposition *to* in the first sentence indicates that *you* is understood as an indirect object. The fact that it is dropped in the second sentence reveals a common English rule of optional elision that has cognitive consequences. When *to* is omitted, there is a greater suggestion that the object was received. Consider, for instance, *John sent Mary the book* as opposed to *John sent the book to Mary*. The first implies that Mary received it, the second does not make so strong a commitment; the book was sent but may not have been delivered.

The result is a question of tone. The first, longer sentence suggests that the communication may have been made without necessarily being registered or accepted by the receiver. This rather subtle distinction is lost in the shorter sentence. That the "to" is missing in "I told you" implies that the receiver *should have* accepted what was said. It creates an evaluative judgment that can elicit an emotional response, reinforced by the exclamation point at the end.

The cognitive principle underlying these two examples is that longer utterances are used when it is necessary to exercise care in communicating exactly what is meant. Long style is therefore more conducive to formal situations such as academic writing or a lecture, when the speaker and audience are physically or psychologically distant from each other. Such situations arise when a reader or audience does not know the writer or speaker very well, and vice versa.

In Dickinson's case, her diction and grammar, despite their seeming idiosyncrasies, share the same linguistic principles of shortened discourse that mark everyday conversational utterances. For instance, a subject in a complement clause can be omitted, as in "We play at Paste - / Till [we are] qualified for Pearl -" (A803, F282B/J320/M531). Repetition of verbs can also be omitted, as in "The one the other will contain / With ease - and [will contain] You - beside -" (A84-5/6, F598/F632/M273). These strategies characterize informal, colloquial speech that indicate intimacy between speaker and hearer. The cognitive rule is thus: "The greater the distance felt between participants, the longer the discourse," and its converse, "the closer participants feel, the shorter the discourse."

In Dickinson's poetry, the most frequently used forms that distinguish informal, conversational utterances from more formal discourse are contractions and elision. Her poems strike an informal note in their spelling and contracting of a morpheme (the smallest meaningful part of a word), as in the following examples:

> Wont you wish you'd spoken (H123, F734/J704/M368)
> 'Twas fighting for his Life he was - (A94-9/10, F1230/J1188/M509)
> And tho' 'tis Years ago - that Day - (H141, F423/J410/M168)
> Good to hide, and hear 'em hunt! (A86-3/4, F945/J842/M441)
> When 'tis small eno' / To credit - 'Tis'nt true - (A84-1/2, F590B/J669/M269)

Repetition of word or phrase is almost always elided, even when the repetition is at some distance, as in the following lines from *It dropped so low in my regard* (H58, F785/ J747/M385):

> I heard it hit the Ground -
> And go to pieces on the
> Stones
> At [the] bottom of my Mind -
>
> Yet [I] blamed the Fate that
> fractured [it] - less

Sometimes, it is a repeated verb phrase or entire predication that is missing:

> Floss wont save you from an Abyss
> But a Rope will - [save you from an Abyss]
>   (F1335/J1322/M582)

Frequently, she will drop a modal or *to* marker before a *be* verb:

> Her pretty Parasol [can / may] be seen (H149,
>   F610/J354/M300)
> Better of it - continual [to] be afraid - (H94,
>   F643/J381/M325)

These are standard ways to shorten informal speech. Dickinson's style is characterized by them, and the resulting tone creates a sense of oral rather than written discourse.

It is not only grammatical components that may fail to appear in a linguistic utterance. Longer discourse allows the writer to spell out situational details that may be left implicit in shorter discourse. Consider, for example, the following sentence included in an email sent to a hiking group:

> Moss behaved well on the hike today.

Presented only with this sentence, a reader could make all kinds of inferences to make sense of its significance. Who is Moss? What hike? What is the presumed significance of mentioning that the hike was "today" or that Moss was well behaved? Recipients of the email did not need to ask these questions because they already knew the answers from knowing the context of the remark. A non-participant needs much more contextual information in order to understand the import of the sentence:

> In our weekly hikes during snow season, I could bring along my hyperactive young border collie, called Moss. Having now been on several hikes, he has learned to settle down when he accompanies us, so that group members can accept his presence more easily.

The less context given, the more opportunities there are for varying interpretations. Meaning resides as much in the nuances of what is not said as in what is. The cognitive rule is that more participation is required in retrieving context when less information is given. Dickinson's poetry causes more reader participation because it uses the following forms that lack context:

1. Unnamed referents
2. Loose appositions/reformulations[2]
3. Ambiguous placements

Sometimes, the referent is never fully identified, as in the poem previously cited, *It dropped so low in my regard*.[3] The reader can only construct what "it" is by recognizing that the speaker has come to blame herself for valuing something that turns out to be worthless. The fact that "it" is never identified generalizes the point of the poem as one of self-blame. In the following poem, the reader must work to identify the actual unnamed referent:

> They ask but our Delight -
> The Darlings of the Soil
> And grant us all their
> Countenance
> For a penurious smile -

> A87-9/10, F908/J868/M428

---

[2] *Apposition* is a grammatical construction in which two elements, normally noun phrases, are placed side by side, with one element serving to identify the other in a different way. *Reformulations* are similar to appositions in formulating ideas in a different way.

[3] Italics are used when citing lines without indicating line breaks, capitalization, and other punctuation markings.

The poem begins with a pronoun, as though the reader already knows who or what is being referred to. The pause after "Delight" invites us to ponder on the range of possible referents for a "they" who don't ask for anything except that we be delighted. The actual subject, placed in apposition in the second line, is not named but described in terms that provide the clue that "they" are flowers and assumes the reader shares the same feeling about them, that they are indeed "Darling." The last three lines deliver the punch that knocks the poem from a silly, sentimental description of flowers into weighted meaning.

The play on "all" in *grant us all their countenance*—does it belong to "us" or "them"?—reinforces our uncertainty. There is a suggestive hint of biblical allusion, with the choice of "grant" instead of "give," and with the long multi-syllabled "Countenance," instead of the more common "face." Even the capitalization contributes: though the flowers ask us to respond with capitalized "Delight" and they themselves give capitalized "Countenance," all we can manage is a lowercase "smile." With the tersely worded contrasts between "grant us all" versus "penurious" and "all their / Countenance" versus "smile," Dickinson elevates the extraordinary bounty flowers give to a level that mocks our inadequate response. The effect of drawing us into the poem as intimate participants makes us feel our own inadequacy at the end.

Even greater participation in supplying the contextual frame is demanded by loose appositions or reformulations. The modifying phrase is so far removed from its head that the reader must supply definitional, interpretive, or elaborative glosses. In some cases, the processing needed is fairly straightforward, as in the following poem, where "A spider and a Flower" are in apposition to "Parties" and "A manse of mechlin and / of Floss (Gloss • sun)" in apposition to "Home":

> The fairest Home I ever
> knew
> was founded in an Hour
> By Parties also that I knew
> A spider and a Flower -
> A manse of mechlin and
> of + Floss   + Gloss • sun
>   A394, F1443/J1423/M605

Reformulations create the possibility for a wide range of interpretations. In the first two stanzas of the following poem, the lines build on the previous ones by the process of reformulation, forcing the reader to reconstruct their syntax (H98, F741/J790/M372):

> Nature - the Gentlest Mother is \
> Impatient of no Child -
> The + feeblest - or the wayward-   + dullest
> est -
> Her Admonition mild -

In Forest - and the Hill -
By Traveller - be heard -
Restraining Rampant Squirrel ₍
Or too impetuous Bird -

The marking at the end of the first line invites us to read the second line as an expanded definition of the word "Gentlest," although one can also read "the Gentlest Mother" as an apposition, with the subject-predicate of the main clause as "Nature is impatient of no child." The phrase, *the feeblest or the waywardest* is also in apposition to *child*, though when we add the next line, "Her Admonition mild -," we are invited to reformulate both lines as appositional elaboration on the first two. But then this reformulation is itself corrected in the second stanza as we must now process "Her Admonition mild -" as subject of "be heard -." By the time we reach the squirrel and the bird, we can see them as examples or exemplars of the children of the first stanza, with the entire sentence starting with "Her Admonition mild -" in apposition to the first three lines. The range of interpretation is wide: Is Nature's lack of impatience with her children the cause or the effect of her being "the Gentlest Mother"? Does the "no Child" encompass the human child too?

The effect of such reformulations is to increase the richness of the possible contexts in which a poem may be read. In the following poem, the final line not only describes the "Undiscovered Continent" but sends out a defiant challenge to the explorer whose goal it is to discover:

Soto! Explore thyself!
Therein Thyself shalt
find
The "Undiscovered
Continent" -
No Settler ₍ had
the mind ₍

                                        A356, F814*B*/J832/M434

The play on the repetition of "thyself" is read first as object of the finding as a continuation of the subject, with object-verb (OV) order: "in thyself [you] will find thyself." However, the next line invites a reformulation: is "thyself" being described as "the 'Undiscovered Continent'" (as an appositional phrase), or is this phrase the actual complement of "shalt find" with "thyself" now to be read as subject? Is the last line to be read as "no settler possessed the mind" or "no mind experienced a settler"? Such reformulations make the reader a participant in discovering possible contextual readings.

Similar to the context/participation principle is the tendency to leave out a clear indication as to the referent of a phrase, or to characterize the subject in general rather than specific terms. This practice occurs when the writer takes for granted the existence of a shared world of knowledge and experience. In less skillful hands, this practice can lead to obscurity, apparent ungrammaticality, or downright incoherence. Some early reviewers of Dickinson's poems criticized them for precisely these "faults":

1. Loss of linguistic marking;
2. Lack of specific reference;
3. Presupposition of shared knowledge or experience.

Phrases such as "Electric Adjunct" (A215, F1424*A*/J1392/M600) or "Germ's Germ" (A90-1/2, F1012/J998/M461) seem to have no reference to "the real world." Sometimes their obscurity lies in the nature of the allusion, as we saw in Chapter 2 with the word *Plush* in *Opon a lilac sea* (A502, F1368/J1337/M590). However, another process is occurring in some of these phrases: their failure to refer to discrete discourse events. Such forms in Dickinson, I suggest, serve to give the illusion of closeness and intimacy.

What is noticeable about certain phrases is that they lack the usual morphological elements that categorize them as signs of particular objects or events. Some of Dickinson's grammatical idiosyncrasies seem clear violations of "well formedness": the lack of mass noun category, for example, in "I wish I were a / Hay -" (H73, F379*B*/J333/M202), or the lack of tense markers on the verb, which seem to give a "subjunctive" or "irrealis" air to the phrase: "A Honey bear away" and "her face be rounder than / the Moon" (H94, F642/J380/M324). Instead of thinking of these as grammatical violations, one can perhaps see them as dependent upon their communicative intent. Their lack of specific reference backgrounds them into the taken-for-granted world of speaker and hearer.

Generalizations are commonplace examples of terms that lack specific reference. The terms *table* and *memory* in the following poem and the capitalized nouns of the second stanza contain no discrete morphological elements that would make them refer to specific things or a particular event:[4]

> One Day is there of
> the Series
> Termed Thanksgiving Day -
> Celebrated part at
> Table
> Part, in Memory -
>
> Neither Patriarch nor
> Pussy
> I dissect the Play -
> Seems it to my
> Hooded thinking
> Reflex Holiday -

[4] I am grateful to Emily Seelbinder for pointing out that Thanksgiving became a national holiday in 1863, during the Civil War (Miller 2016, 773: n. 67).

> Had There been no
> sharp Subtraction
> From the early Sum -
> Not an Acre or a
> Caption
> Where was once a Room -
>
> Not a mention, whose
> small Pebble
> Wrinkled any Sea,
> Unto such, were
> such Assembly
> 'Twere Thanksgiving Day.
>
> A87-5/6, F1110*A*/J814/M495

Here, as can be expected in a longer poem, several conversational techniques are at work that we have seen create a feeling of closeness and intimacy. The theme-rheme reversal, at the outset, of the existential construction "there is one day" accentuates the noun phrase (NP) "One Day" and downplays the message that the speaker is pointing out something new. The omission of the agents for the verbs *termed* and *celebrated* distances the speaker from the actions specified; "I" presumably is not the agent, as indeed is confirmed in the next lines, when the speaker is revealed as reviewer in "I dissect" rather than namer or celebrator. The fact that self-description is given in the negative, "Neither Patriarch nor Pussy" (a variant in another copy is "Neither Ancestor nor Urchin"), suggests that they, unlike the speaker, respectively name and celebrate.

"Seems it," just like the opening phrase, "One Day is there," is syntactically "reflex," backward, topicalizing not the words but their semantic properties of nonspecificity and unreliability, saying in effect to the reader, "Turn me around! And while you are in the habit of turning things around, consider the nature of this 'Reflex Holiday', turn around the negativity of the seemingly chaotic clauses of the following stanza, and it will all make sense."[5]

The long, complex sentence of the second stanza elaborates the meaning of Dickinson's naming Thanksgiving as a "Reflex Holiday" through the cumulative force of the contrasts which are presented in reverse; that is, negatives describe what is, positives what is not, and the elided constructions lend an air of rising emotion or

---

[5] Dickinson's Webster's 1844 dictionary contains the following definitions for the adjective "reflex":

1. Directed back; as, a reflex act of the soul, the turning of the intellectual eye inward upon its own actions—Hale.
2. Designating the parts of a painting illuminated by light reflected from another part of the same picture—Encyc.
3. In botany, bent back; reflected.

Dickinson would also have been aware of current scientific discussions of the "reflex arc," pointing to the turning inward of the new sciences relating mind to matter.

protest at such a state of affairs. The effect is to force us to accede to the statements as incontrovertible facts, so that we, too, at the end, say with the speaker of the poem, "Yes, that really would be an occasion to give thanks!" A full rendition of this stanza, by spelling out the syntax in standard grammatical form and turning it around by unreversing the negatives and positives, would go something like this:

> If the early sum (of what once was, that is, our ancestors) were here, and if there hadn't been a sudden increase in loss of life (possibly a Civil War reference), and if the buildings still existed where there is now only empty space or simply a sign marking the spot, and if the small pebble that was lost at sea was still around instead of simply being talked about (a metaphor for children who have died), then for such an assembly, if it existed, this day would indeed be "Thanksgiving."

One need only contrast the effect produced by my paraphrase—of course, so what else is new?—with the original to recognize how the "reflex" as reflection is given its emotive force, both semantically and syntactically, in referring to what is not there, emphasizing the presence of absence.

   Another common characteristic of Dickinson's practice is the absence altogether of a component part of the discourse—usually the speaker or addressee, who, with the setting, constitute the "ground" of a predication, marked by the following elements:

1. Omission of subjects, agents, and addressees
2. Deixis

These linguistic elements give rise to the following principle: The less specified the ground, the greater the shared perspective.

   When the addressee is not mentioned by name, the effect is a sense of immediacy, of being present in a conversation:

Doubt Me! My Dim Companion!	(H76, F332/J275/M145)
Me, change! Me, alter!	(A281, F281/J268/M531)
Make me a picture of the / sun -	(H78, F239/J188/M105)

Even when the construction used is not an imperative, the effect is still that of direct address:

What if I say I shall / not wait!	(H 69, F305/J277/M159)

Sometimes it is the name of the agent that is missing:

A Shady friend - for Torrid days -	
Is easier to find -	(H69, F306/J278/M160)

In *After great pain* (H26, F372/J341/M198), the agent experiencing the pain is nowhere mentioned: "After great pain, a formal / feeling comes -," and the parts of the body—nerves, heart, feet—are identified by a determiner *the* rather than a possessive pronoun. Similarly, in the last stanza, the agent of the remembering and the outliving is also missing:

> This is the Hour of Lead -
> Remembered, if outlived,
> As Freezing persons, recollect
> the Snow -

The effect of such omissions is to invite listeners to share in the process from their own internal perspective.

The elements that contribute to the gaps, absences, and silences of a Dickinson text thus include contextual knowledge as well as linguistic features: shared assumptions, linguistic gapping, subjective grounding, reformulations, lack of discourse salience. All these elements that share the feature of not stating, of silence, have been shown to reduce the distance between speaker and hearer. In Dickinson's poems, however difficult they may be to process, such strategies serve to increase the intensity of involvement the reader feels.

When the knowledge sets of speaker and hearer are in perfect accord, there is no need for speech: a truism reflected in the common folk notion of an old couple sitting together in absolute silence. Under this view, the language produced is only that required by the needs of the discourse, as in the following example:

> Speech is one symptom of affection
> And Silence one -
> The perfectest communication
> Is heard of none
>
> Exists and it's indorsement
> Is had within -
> Behold said the Apostle
> Yet had not seen!
>   H ST4a, F1694/J1681/M664; lost manuscript

The difficulties of interpretation that a Dickinson text poses arise, ironically, from the amount of knowledge demanded from the reader with respect to a shared cognitive domain. This cognitive domain can be socio-historical, in the sense of allusions to events known by both speaker and hearer, contextual, in presupposing shared referents (as in the biblical reference to the Apostle Peter),[6] or linguistic, in assuming common definitions.

Dickinson's writings are characteristically marked by her use of unidentified and possibly unidentifiable allusions. MacGregor Jenkins (1930, 124f), for instance, in discussing a note Dickinson sent to Jenkins' mother, comments:

---

[6] "Whom having not seen, ye love; in whom, though now you see him not, yet believing, ye rejoice with joy unspeakable and full of glory" (1 Peter 1:8).

I must leave the explanation of this message to others who have more intimate knowledge than I. The words "He was his Country's—She is Time's" may refer to some public event of importance at the time, but it is more likely that it had some intimate, personal meaning clear only to the person to whom it was sent.

In actual discourse, the context of the speech utterance determines what needs to be said and what can be left unsaid. That which can be left unsaid depends on the extent to which the speaker and hearer share knowledge in common. The more intimate the knowledge sets of the participants, the less that needs to be expressed.

Consider, for instance, the following sentence: "I wish I knew just what 'dooms' you meant, though!" So wrote Helen Hunt Jackson to Emily Dickinson, commenting on the last lines Dickinson had excerpted from *Opon a lilac sea* and sent to her as a wedding congratulation. Had she written "what you meant when you used the term 'dooms,'" she would have been asking for a definition of the word which was outside her knowledge set; that is, the distance between speaker and hearer is at its greatest, with zero-shared meanings for the term *dooms*. She could have written, "what you meant by 'dooms.'" In this case, the knowledge sets between speaker and hearer are closer: both have a set of meanings for the term *dooms*; what the questioner is seeking is whether the use of the term has or has not a meaning shared by both speaker and hearer. Helen Hunt Jackson's actual comment, "what 'dooms' you meant," indicates the existence of a shared knowledge set, in which both speaker and hearer share the range of domains for the meaning of the term *dooms*, and the questioner is merely asking the speaker to identify which of the shared meanings the term is invoking. Note that the closer the word comes to the question marker "what," the more salient it appears, and the closer speaker and hearer are.

These distinctions can be formulated as Venn diagrams, where the square = the scope of predication and the circles = the participants' knowledge sets for the term *dooms* (Figure 8.1).[7]

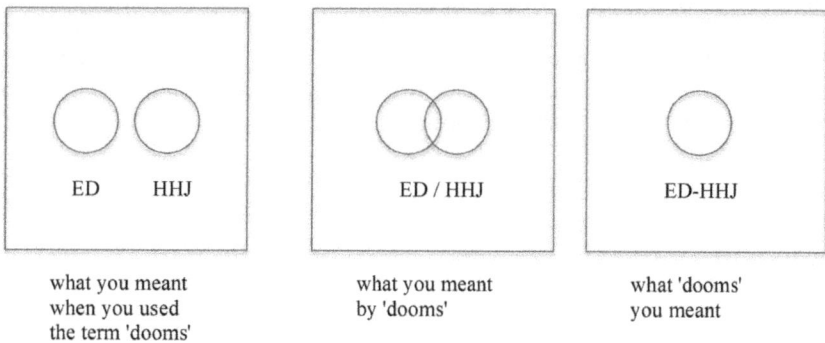

ED HHJ	ED / HHJ	ED-HHJ
what you meant when you used the term 'dooms'	what you meant by 'dooms'	what 'dooms' you meant

Figure 8.1 Scope of predication and knowledge sets.

[7] Named after John Venn (1880), a Venn diagram is a particular kind of way to model information visually. In this case, it depicts the three sentences, with the overlapping spaces indicating

These curious locutions are a part of the cognitive domain that Dickinson demanded her friends and correspondents share with her when they read her poems and letters. As we saw in Chapter 2, one has to work to determine what, exactly, the poem *Opon a lilac sea* is about. One needs to know that Dickinson frequently employs a conceptual metaphor in which SEA represents AIR, that "Balm" can refer to "Beebalm," and that "Plush" is a noun whose referent is a caterpillar.

These principles are so characteristic of the poet's style that they are practically a Dickinsonian signature. It is a cultivation of silence, of not articulating explicitly the grammatical connections, just as the sounds of departure, in their increasing intervals, paradoxically create a greater intimacy through the thoughts they leave behind:

> As Accent fades
> to interval
> With separating
> Friends
> Till what we
> speculate, has been
> And thoughts we
> will not show
> More intimate with
> us become
> Than Persons,
> that we know.
>
> (final lines to *The murmuring of bees has ceased* H345,
> F1142B/J1115/M540)

Assumed in this process is a taken-for-granted world, in which the participants need not express the knowledge they presume to share. The reader of a Dickinson poem must reconstruct its scenario to make explicit the implicit connections of meaning among its words and phrases. As a result, we are drawn in as intimate participants in cognitively experiencing a Dickinson poem.

characteristics the two entities (Dickinson and Jackson) have in common as shared knowledge. Items unique to each entity are represented by the non-overlapping part of their respective shapes.

9

Grounded -Self Spaces

English grammar contains no wildness.

—David Hinton, *Awakened Cosmos*

If, where the rules not far enough extend,
(Since rules were made but to promote their end)
Some lucky licence answer to the full
The intent proposed, that licence is a rule.

—Alexander Pope, *An Essay on Criticism*

When the first edition of a selection of Emily Dickinson's poems appeared in 1890, some literary reviewers criticized her poems for being unmetrical and ungrammatical. Critics today are more tolerant of such poetic violations. Tolerance or intolerance, however, do nothing to illuminate poetic practice. Critics tend to assume that such practices are ad hoc or arise from "poetic license": the belief that the principles underlying poetic language are different from those in conventional usage. I accept neither of these positions. The grammar of language use cannot be determined on the linguistic level alone. It results from both a social/dialectical construct and a personal idiolect. What is needed is a theory that will account for the conceptual models underlying the choices a poet makes in experiencing and conceiving the world.

The grammatical rules that purport to govern the use of pronoun forms in English that carry the -self suffix do not entirely explain why -self pronouns appear when and where they do in actual language use. The failure of both traditional and transformational grammar to fully account for language usage arises from commitment to an objectivist view that grammar generates meaning and that meaning can be characterized by the tools of formal logic. More insidiously, this commitment in itself banishes the imaginative, analogical processes of minding to the realm of fantasy and "untruth," the so-called realm of the poets.

If Dickinson's poetry is considered ungrammatical in its use of -self pronouns, it is so because of the limitations of the grammar, not the limitations of her language. Although Dickinson's grammar is not *prototypical*, it is nevertheless *grammatical*, and the principles of that grammar can be discovered and described. Under this view, poetic license is not freedom *from* the constraints of grammar but freedom *to* construct grammars that conceptualize the poet's worldview. Understanding a poet's grammar can help us understand the poet's worldview and, through it, our own.

According to both traditional grammar and the Government Binding Conditions of transformational generative theory, Dickinson's use of -self pronoun forms seems haphazard and inconsistent. Consider, for instance, the seemingly anomalous use of "Himself himself" in the following poem:

A Spider sewed
at Night
Without a Light
Opon an Arc of
White -

If Ruff it was
of Dame
Or Shroud of Gnome
Himself himself
inform -

Of Immortality
His strategy
Was physiognomy -

H238, F1163A/J1138/M705

In order to account for the use of "Himself himself / inform -," one needs to explore Dickinson's use of the -self pronoun throughout her poetry. For Dickinson, -self pronoun forms in English can be deictic: that is, the characterization of a word or phrase, such as *this/that, now/then, here/there*, that point to the grounding in "mental spaces" of the speaker in a specific time, place, or situation. Analyzed in the light of mental space theory, Dickinson's -self pronouns are perfectly regular.

1 Mental Space Theory

We exist in a world constrained by time and space. We live always in the existential present, at a particular physical location. Thus, at any instant of time we are "grounded" in what we can call our "reality space." The point of view or perspective we take on the world around us is conditioned by the domain structuring that physical space. Such a domain includes our social-cultural knowledge and experiences, our memories, and so on. As humans, we are able to transcend the limits of that reality space by dynamically conceiving other "mental spaces." Gilles Fauconnier (1994) has shown how we dynamically construct these mental spaces in the way we think and reason, and Fauconnier and Turner (2002) have developed the theory further to show how we are able to create new thoughts from these spaces in additional "blended" spaces. That is, we "map" from being grounded in our reality space to other mental spaces.

The term *mental spaces* refers to the fact that, within a single thought or utterance, humans are capable of creating conceptualizations distinct from each other in time, space, or even existence. For instance, when a speaker utters the sentence, "Mary told me she will come tomorrow," the speaker constructs three mental time spaces: the current speech time between speaker and hearer (present); the time of Mary's telling (past); and the prediction of Mary's action of coming (future). A mental space of location is also set up: Mary is currently not in the place occupied by the speaker and hearer (their currently shared reality space) but will be tomorrow. These mental spaces are triggered by explicit or implicit "space builders": here, for example, indicated by the past tense verb phrase, "Mary told." We are also able to create many other kinds of mental spaces: spaces, for instance, that are conjectural, conditional, or counterfactual.

The elegance of the theory of mental spaces lies in its ability to account for a seemingly disparate array of grammatical phenomena involving reference and presupposition without having recourse to rules of a separate and distinct syntax. The power of the theory lies in its ability to shift attention from analysis of the structural complexities of language forms to the mental conceptualizations on which they depend. And the consequences of the theory lie in its ability to explain the capacities of human minding to communicate through conceptual structures: structures that are based, not in analytical reasoning and logic, but in the imaginative, analogical functions of minding, such as metaphor and metonymy, synecdoche and parataxis, parallelism and chiasmus, and all the other figures of rhetoric.

Under this theory, grammatical forms are not simply a matter of syntax or logical relations but arise from the interaction and integration of the ways in which we conceive our experiences.

2 Blending

Whereas earlier metaphor theories introduced two cross-space mappings (variously called "tenor and vehicle" or "source and target"), conceptual integration networks involve at least four spaces (and sometimes more): two input spaces that contribute the elements for the metaphor, a generic space, which shares structure with the input spaces, and a blended space which has "emergent structure," something that exists in none of the other spaces but which emerges from the blend. Although not all blends are metaphorical, all metaphors at some stage in their development involve blending.

The basic blending model has four mental spaces: a "generic" space that identifies structure in common between two input spaces and thus enables mapping between the spaces and projection from them into a fourth, "blended" space (Figure 9.1).

Metaphor occurs when the two input spaces come from different "domains," such as a LIFE domain and a JOURNEY domain that creates the cognitive metaphor LIFE IS A JOURNEY (see Chapter 13). When the structural relationship in the generic space is metaphoric, items from both input spaces are projected into the blend to create

generic space

input space   input space

blended space

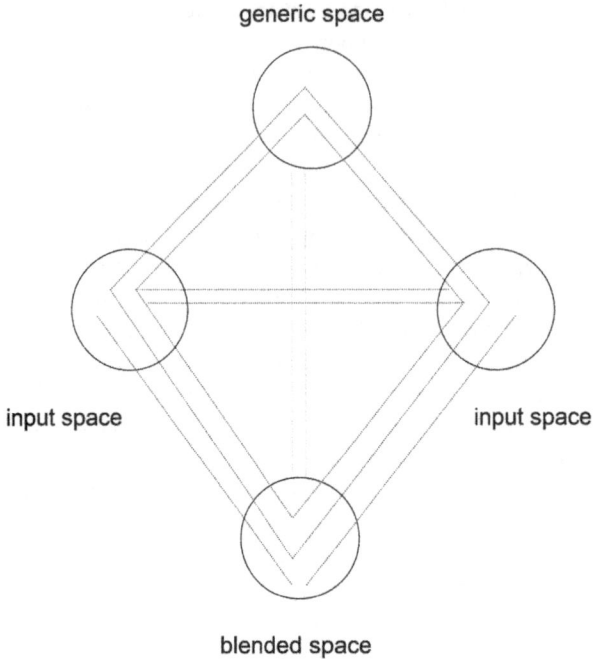

Figure 9.1 Blending model (based on Fauconnier and Turner 2002).

meaning not occurring in either input space (known as double-scope blending). Blending is thus a model for creativity.

The poem as a whole creates what I call a "complex blend." A complex blend refers to the process by which multiple blends create "optimality crossovers" into each other's input spaces when "running" the blend. A poem is also a complex blend in the sense that its possible interpretations are not always immediately apparent; the reader must actively work to understand the nature and relations of its cross-space connections.

3 -Self Pronoun Use in Dickinson's Poetry

In Dickinson's poetry, the identification of counterparts in connected mental spaces forms a complex web of projected -self pronouns and movement between spaces. Dickinson frequently uses regular -self pronoun forms in her poetry:

> You taught me Waiting with Myself -
> Appointment strictly kept -
> You taught Me fortitude of Fate -
> This - also - I have learnt -

                                        H55, F774/J740/M380

> He parts Himself - like Leaves -
> And then - He closes up -
> Then stands opon the Bonnet
> Of Any Buttercup -
>
> H18, F655/J517/M311

Consider, however, the following example (H183, F446/J448/M224):

> We wonder it was not
> Ourselves
> Arrested it - before -

Under any standard account of grammar, such -self pronoun usage appears capriciously irregular, and no purely syntactic theory can explain it. If, however, one looks at these examples in the light of Fauconnier's mental space theory, a different pattern emerges. In the first examples, the pronouns occur within the same mental reality spaces as their antecedents. By contrast, in the second example, the second pronoun reference occurs within a hypothetical mental space that is projected from the speaker's reality space triggered by the space builder "wonder" (Figure 9.2).

The following example is more complex (H180, F845/J920/M422):

> We can but follow to the
> Sun -
> As oft as He go down
> He leave Ourselves a
> Sphere behind -

A different mental space is introduced by means of the space builder "As oft," and within that mental space, the second occurrence of the first-person pronoun "We" takes the -self pronoun form "Ourselves." Since the two references to "He" appear within the same mental space, the -self pronoun form does not occur. However, "He" refers to the "Sun," which *does* occur in the original, speaker's reality space, and could therefore be conceived as a potential antecedent triggering a -self pronoun in the projected mental

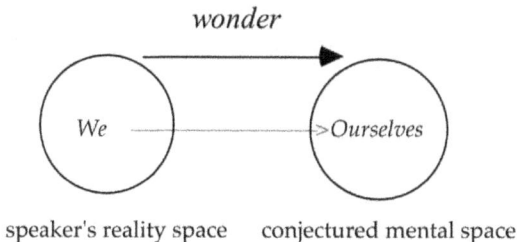

Figure 9.2 "Wonder" as space-building trigger.

space. That it isn't an antecedent for a -self pronoun turns out to be a crucial aspect of Dickinson's use of -self pronouns: they are triggered in mental spaces as projections of *subjects/agents* of their originating spaces. "Sun" in this example is not the subject/agent of the originating space; "We" is, thereby triggering "Ourselves" in the mental space projected through the space builder "As oft." .

I therefore propose the following grammatical rule for Dickinson:

When a subject/agent in one mental space is projected into another mental space via a space-building trigger, its pronoun counterpart in the projected space will take the -self pronoun form.

We can verify Dickinson's rule that the *subject/agent* of the originating space determines the appearance of the -self pronoun in the triggered space. A conditional *as if* space is set up in the following poem in which the sun appears and disappears (H158, F523/J606/M256):

> The Sun shone whole at intervals -
> Then Half - then utter hid -
> As if Himself were optional

A contrastive *but then* space occurs in the following, with the agents in the first space being themselves in another (H91, F658/J538/M313):

> 'Tis True - They shut me
> in the Cold -
> But then - Themselves were
> Warm

-Self pronouns that occur in prepositional phrases are also projected from the subjects of their originating spaces. In the following example, a future time space occurs "in spring," indicating, as the rest of the poem reveals, that the speaker's current reality space is winter (H L17, F4/J5/M698):

> I have a Bird in spring
> Which for myself doth sing -

"It seems" creates a mental space in which "myself" appears (H148, F394/J 588/M210):

> So long I fainted, to myself
> It seemed the common way,

Although the reflexive and emphatic -self pronouns in the following examples could be generated under modified Government Binding Conditions, they also appear as projections of hypothetical and conjectural spaces:

And then I hated
Glory
And wished myself
were They

H195, F1212/J1227/M503

I $^+$ could have done a Sin
And been Myself that
$^+$ easy Thing
An independant Man -
+ might +distant

H162, F856/J801/M393

In the following example, two pronominal subjects—"myself" and "He"—occur in the projected "Uncertain if" space. The word "pain" in the originating space is co-referential with "He" in the projected space, but does not project a -self pronoun because it is not a subject in the originating space, whereas "I" is (H139, F288/J574/M296):

A'blossom just when I went in
To take my $^+$ Chance with pain -   + Risk
Uncertain if myself, or He,
Should prove the $^+$ strongest One.   + supplest

This subject constraint does not preclude double mental spaces with two subjects, as in the following example (H201, F199/J207/M118; **boldface** used to indicate the second co-referential subjects):

To think just how the fire will
burn -
Just how **long-cheated eyes** will turn -
To wonder what myself will say,
And what **itself**, will say to me -

Here the speaker is projecting a fantasy of arriving home late. "I" is the underlying subject of the space builder "To think," triggering a mental space in which "long-cheated eyes" is a subject. The second space builder, "to wonder," can be read as doubly triggered, from the initial "I" of the speaker, but also from the "long-cheated eyes," so that two subjects occur with -self pronouns in the resulting mental space. Dickinson's use of the -self pronoun might also illumine the question of whether reflexive pronouns, like emphatics, are generated by cognitive rather than syntactic constraints, such as the plural "long-cheated eyes" as metonymy for a person in the previous example equating with singular "itself."

Some examples of pronoun usage in mental space projections in Dickinson's poetry appear as exceptions or counterevidence to my proposed principle. In the following, the -self pronoun reference is more complex, occurring at a distance across separate sentences and stanzas (H182, F445/J538/M223):

> They shut me up in Prose -
> As when a little Girl
> They put me in the Closet -
> Because they liked me "still" -
>
> Still! Could themselves have peeped -
> And seen my Brain - go round -
> They might as well have lodged
> a Bird
> For Treason - in the Pound -

According to the -self pronoun principle, the "As when" creates a time different from the opening line, so one should expect the agent "They" to take the -self pronoun form in the first stanza within that time space. But it doesn't. What this means, I think, is that the lines "As when a little Girl / They put me in the Closet - / Because they liked me 'still' -" are operating as an additional analogical mental space, an interpolation rather than being triggered from the first line. The -self pronoun "themselves" does occur as expected in the second stanza, with the space builder "Could" triggering a counterfactual mental space from the statement "They shut me up in Prose -." What this suggests is that sometimes it is the subject/agent of the entire poem that governs -self pronoun use.

In the following poem, *Heaven has different signs to me* (H140, F544/J575/M297), the appearance of "itself" might indicate that -self pronouns are not always projected from the grammatical subject of the parent space:

> All these - remind us of the
> place
> That Men call "Paradise" –
>
> Itself be fairer - we suppose -
> But how Ourself, shall be
> Adorned, for a Superior Grace -
> Not yet, our eyes can see -

The space builder "remind" creates a hypothetical space in which "all these" is equated with "the place" called "Paradise." "All these" refers to the signs that have been described in the previous stanzas that stand for the poem's subject: heaven. The projection from space to space allows noun phrases like "All these," "the place," "Paradise," and "Itself" to be linked to the subject/agent "Heaven" of the poem's first line. The reflexive thus can be licensed by one form (All these) and agree grammatically with another (Heaven → the place → Paradise → itself).

Dickinson habitually turns her sentences around. The last stanza can be paraphrased: "We suppose itself be fairer. But our eyes cannot yet see how ourself shall be adorned for a superior grace." As we have seen, the -self pronoun in "Itself be fairer" is projected from the *subject* of the originating space, "All these," making "we suppose" an appositive space builder, or a multiple connector in Fauconnier's terminology. Set up in "But" contrast to this space/place called "Paradise" is the future negative space projected by "Not yet, our eyes can see -" in which "our eyes," standing for "I," projects the -self pronoun "Ourself" in that space.

When mental spaces are multiply embedded, the subject in one space will project its own -self pronoun into the mental space projected from it. Thus, in the following example, the speaker is explaining why she "never felt at Home - Below -" and projects a conditional space in the third stanza:

> I never felt at Home - Below -
> And in the handsome skies
> I shall not feel at home
> I know -
> I dont like Paradise -
>
> Because it's Sunday - all the time -
> And Recess - never comes -
> And Eden'll be so lonesome
> Bright Wednesday Afternoons –
>
> If God could make a visit -
> Or ever took a Nap -
> So not to see us - but they
> say
> Himself - a Telescope
>
> Perennial beholds us -
> Myself would run away
> From Him - and Holy Ghost - and All -
>
> H143, F437B/J413/M175

In the conditional *if* space in which "God" is a subject, a contrastive space is created with the words "but they say," and the subject "God" projects onto the third-person -self pronoun in the embedded mental space of "Himself - a Telescope / Perennial beholds us -." It is not clear, however, why "Myself" instead of "I" occurs in the following line. However, the first stanza sets up the grounded spaces of "I" as subject/ agent governing the entire poem. The second stanza acts as a pivot where the "because" clause gives the reason why "I dont like Paradise - " and why "Myself would run away." This syntactical looking both ways is characteristic of Dickinson and creates a doubling effect so that "Myself would run away" is both the result of the speaker's feelings and an opportunistic expression of what would be possible if only God would absent himself for a while. An even more complex example of Dickinson's -self pronoun usage occurs with what I call "crossover spaces."

4 Crossover Spaces

As multiple mental spaces are generated, the perspective or point of view can shift from one space to another. An example can be seen in James McCawley's (1981, 328) famous sentence: "I dreamt that I was Brigitte Bardot and that I kissed me." In any traditional view of grammar, the structure noun phrase—verb phrase—noun phrase (NP VP NP) would produce a reflexive pronoun in the second NP, as in "Harry hurt himself." However, in the dream space, the -self pronoun rule is blocked by cross-space identity connectors that link "I" to Bardot and "me" to McCawley. Since "I" now deictically refers to Bardot in the dream space and not to McCawley in the originating space, the pronoun referring to McCawley is "me," not "myself." This ability for identity to cross spaces within a single utterance is, as we shall see, characteristic of Dickinson's more complex use of the -self pronoun forms.

In Dickinson's poetry, the form of a pronoun reference will shift as the perspective shifts among the mental spaces that are projected, and pronominal references *cross over* into other spaces. Consider, for example, the last two stanzas of *I tried to think a lonelier thing* (H85/384, F570/J532/M260):

> I plucked at our Partition -
> As One should pry the Walls -
> Between Himself - and Horror's
> Twin -
> Within Opposing Cells -
>
> I almost strove to clasp
> his Hand,
> Such Luxury - it grew -
> That as Myself - could pity
> Him -
> + Perhaps he - pitied me -   + He - too - could pity me -

The speaker fantasizes a mental space in which she creates an entire scenario that there might be someone else besides herself "Of Heavenly Love - forgot -." The -self pronoun "Himself" occurs regularly, following "One" within the mental space triggered by "As." In the last stanza, the speaker considers reaching out to "clasp / his Hand," and in that mental space of projected fantasy refers to herself as "Myself" as expected. The last line seems to contradict the -self pronoun rule, since the subject/agent of the entire sentence is "I" so that "myself" should occur in both -self references in the projected mental space. However, whereas she, as "Myself," can pity him in that mental space she has fantasized, he might pity her, not in that same fantasized space, but in her own reality space. The crossing of mental spaces is what causes the effectiveness and power of the ending, as can be tested by substituting "myself" in that final line: "+ Perhaps he - pitied myself - (+ He - too - could pity myself -)." The result of such a substitution is to stay in the same mental space of fantasy; there would be no crossing of spaces into the speaker's reality space, and the power of the ending would be lost.

The practice of crossover spaces seems to indicate that the -self pronoun grounds the self *physically* in the mental space into which it is projected. The following discussion shows that this is in fact the case.

Several of Dickinson's poems deal with the existence of presence in death, thus complicating the nature of the "reality" spaces being presented. For instance, the poem *'Twas just this time last year I died* (H61, F344/J445/M181) is a narrative, being told from the perspective of death. Through the space builders, *I know, I thought, I wondered,* in the first five stanzas, the speaker in the grave projects the events of real life that are going on without her. Had Dickinson used the -self pronoun in these stanzas, she would have been deictically projecting the speaker into that mental space of life, thus physically "grounding" the speaker in the objective scene. However, the very point of the poem is that the speaker is physically absent from the mentally projected scene of real life.

In these stanzas, then, when she refers to herself, the speaker crosses over from the mental space of life to her reality space of death so that the pronoun does not take the -self form:

> I wondered which would
> miss me, least,
> And when Thanksgiving, came,
> If Father'd multiply the plates -
> To make an even Sum -
>
> And would it blur the
> Christmas glee
> My Stocking hang too high
> For any Santa Claus to reach
> The Altitude of me -

The last stanza of the poem contrasts the mental spaces of the entire poem, between life and death:

> But this sort, grieved myself,
> And so, I thought the other
> way,
> How just this time, some
> perfect year -
> Themself, should come to me -

Now the -self pronoun occurs, as thinking about life creates a contrastive *but* mental space in which the speaker projects her grief. The poem ends in yet another projected space, "some perfect year," and in this "other way" space, this time the -self form is attributed to the subject/agents in that other space in life who will come to the speaker's reality space in heaven: "Themself, should come to me -."

-Self pronouns can thus appear or not appear in projected mental spaces when crossovers occur with shifts in focus, perspective, and point of view. Seemingly inconsistent use of -self pronouns is thereby resolved and explained. Another poem with "crossover spaces" creates mental spaces for both the dead and the living:

> Those + fair - fictitious People -
> The Women - + plucked away
> From our familiar + Lifetime -
> The Men of Ivory -
>
> Those Boys and Girls, in Canvas
> Who + stay opon the Wall   + dwell
> In Everlasting + Keepsake -
> + Can Anybody tell?
>
> We trust - in places
> perfecter -
> Inheriting Delight
> + Beyond our + faint Conjecture -
> Our + dizzy Estimate -
>   + small   + scanty
>
> Remembering ourselves - we trust -
> Yet Blesseder - than we -
> Through Knowing - where we
> only + hope
> + Receiving - where we - pray -
>
> Of Expectation - also -
> Anticipating us
> With transport - that would
> be a pain
> Except for Holiness -
>
> Esteeming us - as Exile -
> Themself - admitted Home -
> Through + gentle Miracle of
> Death -
> The Way ourself - must come -
> + new   + address - gazing - + guess + beholding +
> curious - <u>easy</u> -

+ slipped away - + familiar notice + [familiar] fingers +
Everlasting Childhood -
Where are they - Can you tell -

H25, F369/J499/M196

Three lines with parallel surface forms appear to be inconsistent in pronoun use:

Remembering ourselves (stanza 4)
Anticipating us (stanza 5)
Esteeming us (stanza 6)

The speaker, looking at representations of the dead, and wondering where they are, trusts they are "in places perfecter" (stanzas 1–3). The next stanza projects into the space of the dead from the speaker's reality space through the space builder "we trust," triggering as expected the -self pronoun in the projected mental space as the dead become the subject of that space ("we trust they remember"): "Remembering ourselves, we trust" (Figure 9.3).

Starting with the next line, the *perspective* shifts from the reality space of the speaker to that of the dead through the space-building comparative terms "yet" and "where," contrasting the state of the dead in their space to that of the living in the originating (speaker's reality) space:

Yet Blesseder - than we -
Through Knowing - where we
only hope -

Although the dead "know" in their domain where we can "only hope" in ours, they also experience, as we do, "Expectation":

Of Expectation - also
Anticipating us

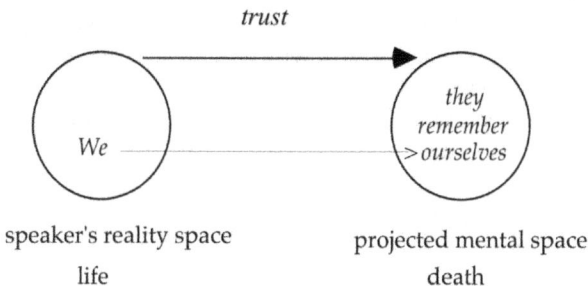

Figure 9.3 "Trust" as space-building trigger.

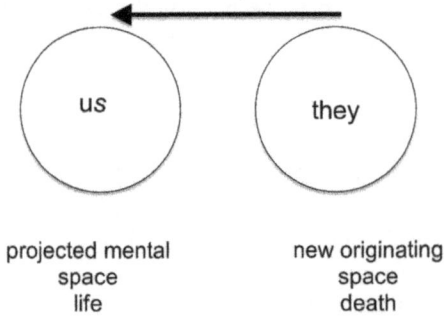

Figure 9.4 Space projection from death to life.

As the poem shifts into the mental space of the dead as the *originating* space, the subject of that space is "the dead" not "we," so the pronoun in the now projected space of the living takes the regular, not the -self pronoun form (Figure 9.4).

Now it is the speaker's space that is "away" in "Exile," and the final stanza ends in the domain of the dead:

> Esteeming us - as Exile -
>
> Themself - admitted Home -
> Through gentle Miracle of Death -
> The way ourself, must come -

With the deictic words *must come*, the pronoun *ourself* is being projected from the perspective of the mental space of the dead as is *themselves*, with the word *admitted*. It is, however, still governed by the "parent" space of life now, and the -self pronoun occurs only when the subject self in the mental space of the living is projected into the mental space of the dead.

By projecting the self from life now into the world of the dead, whether as the speaker's reality space in "Themself, should come to me -" in the previous poem, or the mental space of the dead in "The way ourself, must come -," Dickinson makes the world of the dead an integral part of our own (Figure 9.5).

While the argument might be made that a -self pronoun projection into the ground of another mental space is *physically* deictic by definition, the stronger argument comes from the poet herself. Following are two poems Emily Dickinson sent with flowers. In the first poem, the -self pronoun *themself* that opens the poem refers to the "all" and literally points outside the mental space of the poem to the physically grounded space of the accompanying bouquet:

> Themself are all I
> have -
> Myself a freckled - be -
> I thought you'd choose
> A Velvet Cheek

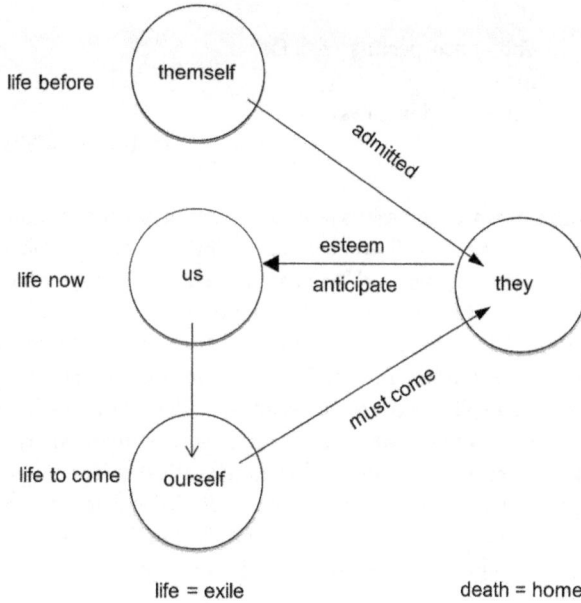

Figure 9.5 Space projections from life to death.

> Or one of Ivory -
> Would you - instead
> of Me?

H193, F1054/J1094/M473

The subject "I" triggers "Myself" in the contrasting space that is set up, with the NP-be-NP construction. Another mental space is set up with the space builder "I thought," as the speaker contemplates the flowers' recipient choosing among them. The final question, like the crossover examples previously discussed, returns to the speaker's reality space, as she phrases the question directly, and, as expected, "Me" appears instead of the -self pronoun form.

On at least one occasion, Dickinson literally tucked a small note inside the flower she sent:[1]

> I hide myself - within
> my flower,
> That fading from your

---

[1]   Several versions of this poem exist in manuscript form. T. H. Johnson thinks it possible that other copies were made. He quotes Mary Adèle Allen (*Around a Village Green*, 1939) who "says that ED once sent a bouquet to her mother and that one flower was bent back and a tiny note placed in it. 'The note has long been lost, but I wonder if it might have been the poem in which she hides herself within her flower'" (Johnson 1955, 664–5).

> Vase -
> You - unsuspecting - feel for
> me -
> Almost - a loneliness -
>
>   H210, F80C/J903/M56/M408

Here, the -self pronouns are themselves multiple, equating poet, poem, and flower. The future mental space image of the flower containing "myself" fading evokes the possible response of the recipient who will feel "Almost - a loneliness -," not for the poet-poem-flower in the fantasy space but the poet in her reality space.

On one occasion, Dickinson herself had second thoughts about what form to use. As native speakers, we develop an intuitive ability to use the dialectal grammar of our milieu without an explicitly conscious awareness of its workings. That is, we don't know what it is we know when we speak (unless we are linguists). That Dickinson had an implicit, idiolectal knowledge of her grammatical rule is not surprising. In the following poem, her implicit -self pronoun rule led her to use "myself" when she wrote out the poem, *I took my power in my hand* (H91, F660/J540/M313) in which the speaker compares herself with David. The -self pronoun occurs twice in the second stanza:

> I aimed my Pebble - but
> Myself
> Was all the One that fell -
> Was it Goliah - was too
> large -
> Or + was myself - too small?
> + just myself - Only me - I -

The variant for the last line (marked with a cross in the manuscript) indicates some uncertainty on Dickinson's part as to which pronoun form to use. The first use of "Myself" in this stanza is predicted by the projection of the subject "I" into the contrasted mental space set up by "but." The second "myself" is interesting. It occurs in a mental space set up by the question "Was it," with "Goliah" as the subject of that originating space, and a subsequent alternative *or* space in which "myself" should not be triggered. However, if both spaces are multiply projected from the first sentence of the stanza, "I aimed my Pebble -," then "myself" is generated by Dickinson's grammatical rule.

Dickinson rarely alters the metrical rhythm of her variants. The possible preference for "Only" over "just" in her suggested variant for the last line may have led Dickinson to use the monosyllables "me" or "I" to keep the metrical rhythm. It is, of course, impossible to know precisely what Dickinson was thinking in her suggested variants, but this example raises interesting questions about the underlying affective forces motivating her choices.

One remarkable use of -self pronouns in crossover spaces is found in the following poem, in which the speaker splits the self into two parts:

> Me from Myself - to banish -
> Had I Art -
> ⁺ Invincible my Fortress
> ⁺ Unto All Heart -
>
> But since Myself - assault Me -
> How have I peace
> Except by subjugating
> Consciousness?
>
> And since We're mutual
> Monarch
> How this be
> Except by Abdication -
> Me - of Me -?
> + impregnable + To foreign Heart -

H42, F709/J642/M345

The pronoun usage in this poem can be explained through the extension of mental space mapping into a complex blending account. The dominant images of the source space include a ruler (monarch), a subject, a fortress, and the idea of banishment (stanza 1), assault by an enemy and the idea of subjugation (stanza 2), and a monarch and the idea of abdication (stanza 3). These are all metaphorically projected onto the target space which is the self. What is happening in this poem is similar to Fauconnier and Turner's (1994) account of Dante's description of Bertrand de Born in the *Inferno*. By dividing the self into two parts, a blended space is projected which is "impossibly in conflict with our understanding of actual human beings" (5).

In Dickinson's poem, the pronoun forms *me* and *myself* are being used as cognitive projections of "I," which is the self, itself, and the metaphors of monarch, subject, fortress, and war are projected onto the self to create the argument of the poem (Figure 9.6).

Within the blended spaces of the first two stanzas, the self-role "I" takes on two values, "myself" and "me." In the first stanza, "Had I" is a space builder, triggering the expected -self pronoun "Myself" in the projected mental space. To banish someone carries the associated images of place: the one who banishes drives someone else from the place one is in or occupies. The -self pronoun is attached, not to the one being banished (subject-heart-me), but to the reference most closely associated with the "Fortress" that the self commands (monarch-mind-myself). In the second stanza, a contrastive mental space is set up by the space builder "But since," and the -self pronoun (subject-mind-myself) is this time assaulting "Me" (monarch-

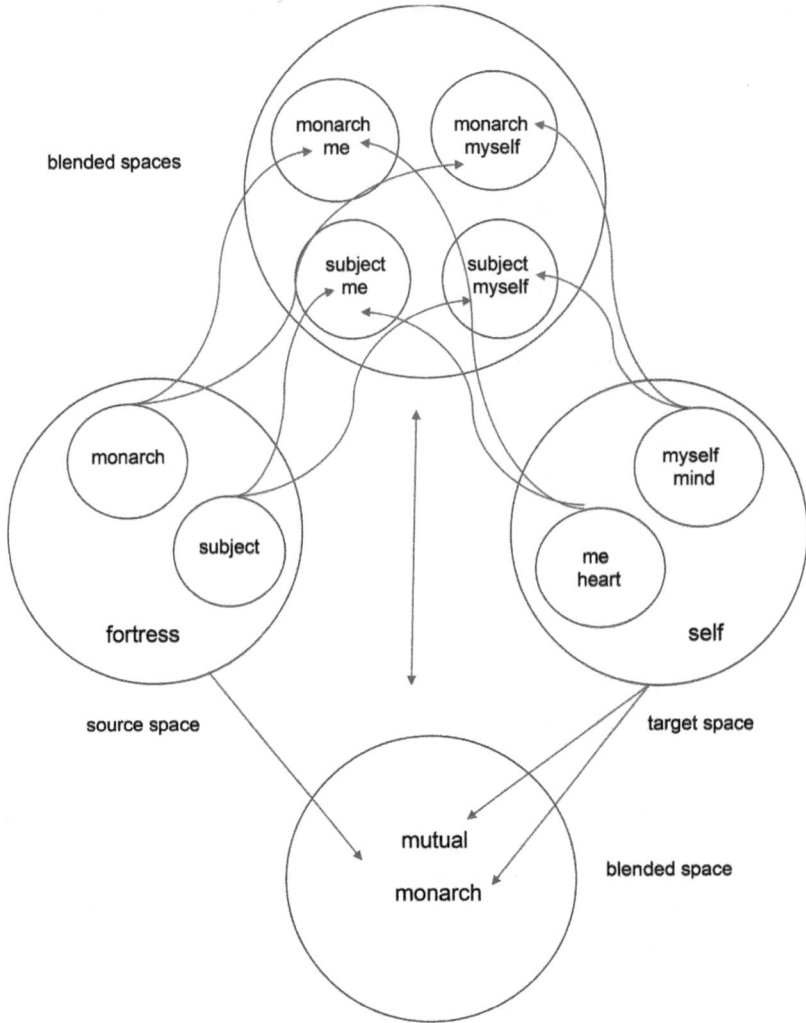

blended spaces

monarch
me

monarch
myself

subject
me

subject
myself

monarch

myself
mind

subject

me
heart

fortress

self

source space

target space

mutual

monarch

blended space

Figure 9.6 -Self pronoun use in complex blending.

heart-me), so that the self-role, "I," must step in to "subjugate." The third stanza explodes the fantasy of these stanzas by recognizing that "We" (both "Myself" and "Me") are "mutual Monarch," so that the self cannot in fact banish or subjugate but only abdicate. To "abdicate," however, like "banish" in the first stanza, is to abdicate from something (the throne) that is occupied by the abdicator ("Monarch"), so that the self-role takes on the value of place. But since the place *is* the self, the question arises, "How this be," since "Me" *is* "Me." A reading of the argument of the stanzas, as the images of banishment, war, and abdication are projected from the source space and the me/myself pronouns are projected from the target space into the blended space of fantasy, goes something like this:

> Stanza 1: If I were able to suppress my emotion (heart-me), then I would be fully protected in the fortress of "Myself."
>
> Stanza 2: But since it is my mind (consciousness-myself) that is assaulting my heart-me, then I would need to suppress my consciousness myself in order to have peace.
>
> Stanza 3: Because the mind and the heart both rule the fortress of the self and are united as one, the only way it seems I can resolve the problem is to give up sovereignty of self altogether.

That is, the three stanzas form a three-part argument:

> I don't have the art/power to banish my heart-me,
> Therefore, I will have to subjugate my consciousness-myself.
> But since I (myself and me) am all there is, the only way this could work
> would be to abdicate from the place of my self, "Me - of Me -."

The first two stanzas set up the fantasy of the divided self, between heart and mind. The last stanza, however, reinforces the integrity of the self, "We're mutual / Monarch," and questions the impossibility of the fantasy, "How this be." Had the last line maintained the fiction of the division between "myself" and "me," the original metaphors would have remained dominant within the blended space. By repeating the same form, "Me - of Me -?" the target of the unified self is emphasized. The point of view has switched from the blended spaces, in which the -self pronouns predictably occur, to the target, "reality" space of the unified self. In other words, the final stanza has returned or crossed over into the speaker's reality space of "Had I Art" of the first stanza. In returning to the reality space of the speaker's "I," Dickinson reverts to the regular "me" form and the -self pronoun disappears.

5 Cognitive versus Grammatical Analysis

Grammatical analysis alone cannot account for Dickinson's use of the -self pronoun. Relying upon generative grammar, as the Tübingen linguists (Bauer et al. 2020) do in their analyses, unnecessarily complicates possible readings. For instance, in analyzing Dickinson's poem, *This was a poet* (H183, F446/J448/M224), they comment on the last stanza "Of Portion - so unconscious - / The Robbing - could not harm - / Himself - to Him a Fortune - / Exterior to Time -," as follows:

> We will exclude the possibility that "himself" is the argument for "harm." According to our two analyses of the sentence, neither of them provides a basis where it is grammatical to use "himself" as an argument for "harm." In the reading where the

speakers rob the poet without harming him, the reflexivity would not make any sense as the poet is not the agent of the robbing event; the other reading would fully account for the reflexivity of the pronoun, and, given that if the poet robs somebody else, it is less plausible (but possible) that this action leads to him harming himself. Since the first reading suggests that it is not the poet who is responsible for the robbing but the speakers, using a reflexive is dispreferred. (68, n. 12)

By excluding the possibility of "himself" being the object of "harm," the linguists are forced to miss the syntactic ambiguity of "Himself - to Him - a Fortune -" and thus to read the line as stating "He is a timeless fortune for himself," instead of seeing the "fortune" as the gift of poetry. A cognitive reading both simplifies and clarifies the final stanza.

> This was a Poet -
> It is That
> Distills amazing sense
> From Ordinary Meanings -
> And Attar so immense
>
> From the familiar species
> That perished by the Door -
> We wonder it was not
> Ourselves
> Arrested it - before -
>
> Of Pictures - the Discloser -
> The Poet - it is He -
> Entitles us by Contrast -
> To ceaseless Poverty -
>
> Of Portion - so unconscious -
> The Robbing - could not harm -
> Himself - to Him - a Fortune -
> Exterior - to Time -

H183, F446/J448/M224

The entire poem hinges on and follows from the opening statement "This was a Poet." The first two stanzas compare what the poet does with what we as readers don't: we can only wonder at the power of the poet to conjure the amazing sensory perception (here of smell) from the experience of pressing attar from a rose. If "That" in the second line is understood as a relative pronoun with elision of the poet (It is he that . . .), it introduces the parallel repetition of "The Poet - it is He -" in the third stanza.[2] The last

---

[2] Recognizing the possibility of *that* as a relative pronoun, though suggesting it could refer to the poem, not the poet, Bauer et al. (2020, 55: n. 3) prefer to "simplify by not including this reading in the discussion," thereby introducing unnecessary complication into their reading.

two stanzas thus amplify and comment on the contrast set up in the first two. Just as we cannot lay claim to the poet's power in the first part, we are entitled ("give a claim to, to demand or receive" in Webster's definition) to "ceaseless Poverty" in contrast to the poet's "Fortune." Just as the poet is the discloser of "Pictures" in the third stanza, so is he unconscious of "Portion" in the last.

Because the linguists cannot accept the grammaticality of "Himself" as an object of "harm," they miss the climactic point of the last two lines. Because it is the poet who is the subject/agent of being unconscious of portion, reference to the poet in the projected space of possibility, according to Dickinson's -self pronoun principle, becomes the reflexive "Himself." Just as it is the poet, not the reader, who does the "arresting" in the first part, it is the reader, not the poet, who does the "robbing" in the second. To the poet, the "portion" acquired by the power of poetry is so great that it is a "Fortune" that, conforming to the ideas of the Romantic poets, lasts beyond the boundaries of human time.

By using the principle of -self pronoun projection from the subject/agent in one mental space into another, Dickinson creates for us a *world of possibilities*: a world in which things can happen and be made to happen through the agencies of the self. Whereas linguists have expertise in how language works, it is the poet who has mastery of making language work.

Dickinson's Spider poem quoted at the beginning of this chapter is one of those cases of ungrammaticality people characteristically point to. The seemingly anomalous use of the double reflexive in "Himself himself inform" can be explained by Dickinson's -self pronoun usage. In the poem, the subject "Spider" projects onto the -self pronoun subject of the conditional space set up by the *if* space builder. The reflexive object is then preposed before the verb to get the line "Himself himself inform."[3]

But what are the consequences of Dickinson's manipulation of the -self forms in this as in other poems? The spider builds his web at night. Only he can give shape to or "in-form" the nature of his projection of himself into his web as he builds it. The grounding of the self in the space of the web is both strategic and physical: whether it be the accroutrements of life (the "ruff of dame") or the accroutrements of death ("the shroud of gnome"), the spider is building his immortality. But it has to occur through a *physical* projection of himself into his work—the web, his physiognomy.

Such an achievement can only occur through the spider physically projecting himself into his embodied space of being. We need to project self into world to change the world, to give it meaning, to create the web. If the spider's web is to have any meaning, then the spider must create itself in a mental space: it must project itself as an objective presence in the world. This poem is not simply about poetry, it is a poem about how poetry works in the world.

---

[3] Dickinson's practice of preposing objects before the verb from normal subject-verb-object (SVO) structure sometimes creates ambiguity as to whether the sentence is SOV or OSV. Here of course, the line can be read either way.

The Presence of Self

Emily Dickinson understood that not only human beings have selves but all creation. Her poems scintillate with the possibilities of reciprocal, antiphonic relationship, not only with others but with the natural world of animals, plants, weather, death, life, and time itself. In addition, the rise of the dramatic monologue in lyric, influenced most notably by Robert and Elizabeth Barrett Browning, crept into her poetry as she adopted many different roles and voices (Finnerty 2014). Dickinson expands these practices to include all the beings and events of nature. To experience a Dickinson poem, one needs to recognize cognitively the various selves that are present.

In this chapter, I explore some of the ways Dickinson engages with self and other in developing the themes of her poems. Both antiphonal play and the dramatic monologue enable role-playing: the adoption of stances to take up a particular attitude within the context of the situation being addressed. Each individual self has many roles depending on many factors. For instance, I am an educator and a writer, a wife, a sister, a hiker, a gardener, a friend, a neighbor, and so on. A poem's content and context guide the reader in identifying the kind of self being portrayed, either by third-person description or through an "I" persona.

Dickinson is not a "confessional poet" as other poets have identified themselves, mainly during the late 1950s and early 1960s—poets such as Robert Lowell, Sylvia Plath, and Anne Sexton—where use of "I" consciously reflects actual details of the poet's own personal life. The perspective of "voice" we hear in Dickinson's poems does not always represent Dickinson herself personally, though sometimes it does. In her poems, Dickinson creates a "poetic self" to distinguish the identity of a narrative voice from her personal self or to introduce the perspective of another. The voice of the poetic self distances itself from personal anecdote, just as my "voice" in this book distances the writing self from my other self-roles. This is not, of course, to deny the obvious fact that the motivating force of Dickinson's poems comes from her.

In this chapter, therefore, I explore the presence of various selves as they appear in Dickinson's poems, while recognizing always the motivating voice of Dickinson herself behind all poetic utterances.

1 The Selves of Nature

Dickinson speaks in her poems with many voices and adopts many roles, both human and natural. One's experience of self presented by the I-persona of a poem is determined by the establishment of a context and identity, as in the identification "And I'm a Rose!" in *A sepal petal and a thorn* (A82-1/2, F25/J19/M34) previously quoted. Sometimes, only the context and content provide the identification of a poem's subject:

> There is a flower that
> Bees prefer -
> And Butterflies - desire -
> To gain the Purple Democrat
> The Humming Bird - aspire -
>
> And Whatsoever Insect pass -
> A Honey bear away
> Proportioned to his several
> dearth
> And her - capacity -
>
> Her face be rounder than
> the Moon
> and ruddier than the Gown
> Of Orchis in the Pasture -
> Of Rhododendron - worn -
>
> She doth not wait for June -
> Before the World be Green -
> Her sturdy little Countenance
> Against the Wind - be seen -
>
> Contending with the Grass -
> Near Kinsman to Herself -
> For privilege of Sod and Sun -
> Sweet Litigants for Life -
>
> And when the Hills be full -
> And newer fashions blow -
> Doth not retract a single
> spice
> For pang of jealousy -
>
> Her Public - be the Noon -
> Her Providence - the Sun -

Her Progress - by the Bee -
proclaimed -
In sovreign - Swerveless Tune -

The Bravest - of the Host -
Surrendering - the last -
Nor even of Defeat - aware -
When cancelled by the Frost -

H94, F642/J380/M324

Dickinson invokes the life of a clover as it interacts with its environment: offering nectar to birds, bees, butterflies, and other insects; contending with the grass for sustenance; indifferent to the more glamorous plants of summer; and in its bravery "nor even of Defeat - aware - / When cancelled by the Frost -." By using "Purple Democrat" to describe the lowly clover, the speaker not only refers to its color, but to the trappings of royalty, suggesting that even the lowliest has value. In the clover's Darwinian fight for survival with the grass, Dickinson extols not just the flower but a principle of democracy in both being "Sweet Litigants for Life."

By identifying characteristics of natural beings and plants, Dickinson is not committing the pathetic fallacy of anthropomorphizing nature. Rather, she is rejecting the notion that basic cognitive processes are what distinguish human beings from the rest of nature. Close attention to natural forces enables her to see that bees, butterflies, and hummingbirds do indeed prefer, desire, and aspire to some flowers over others. Frequently, Dickinson creates a reciprocal relationship with the many selves of others and nature. Squirrels have their own "estimate" (A110, F1407/J1374/M595); a bat has his own "eccentricities" (F1408/J1575/M596). Sometimes, she assumes a plural identity, as in *Ourselves we do inter* (A706, F1449/J1144/M718); most often the reference is singular.

Even immaterial experience offers up the possibility of interaction between self and other. In the following poem, Death is understood as a trade exchange:

For Death - or rather
For the things 'twould buy -
This - put away
Life's Opportunity -

The Things that Death will
buy
Are Room -
Escape from Circumstance -
And a Name -

With Gifts of Life
How Death's Gifts may Compare -

> We know not -
> For the Rates - lie Here -

H 94, F644B/J382/M325

The idea of the gift of life is a common enough expression, as is the contrast between life and death. What Dickinson does is to conceive of the contrast in equitable terms. If life has gifts, then death must have them too. The presence of self in this poem is complex: Life and Death can be gift givers, but they can also be commodities or transactions—money or things to trade. With the additional imperative "put away," an addressee is invoked, followed by the "we" of the final stanza. So that comparison is ascribed to the human self, but a self that cannot know death's "Rates." The enigmatic ending causes us to ask "Where?" By turning back into the poem, we see that the somewhat puzzling addition of the grammatically unnecessary "This -" of the first stanza, explained by "Life's Opportunity," anticipates the "Here -" of the last. If we "put away"—ignore—the opportunity life gives, then what death offers does indeed "lie" here.

From natural events and processes, the four elements of earth, air, water, and fire to inanimate rocks and mountains, to plants and trees, to animals and finally to human beings themselves, Dickinson inspires all with agency.

2 Antiphonal Play

The word *antiphony* comes from the Greek ἀντίφωνον, ἀντί "opposite" and φωνή "voice." It characterizes responses between two groups or voices, and is related to dialogue, a conversation between two or more people. Dickinson was well acquainted with the practice of antiphonal singing in the Congregational church her family attended, both with pastor/congregational responses and the sharing of hymn verses between male and female voices as indicated in her Worcester-Watts hymnal. Antiphonal play underlies Dickinson's dialogic presentations of alternative selves.

The following poem presents a dialogue between a brook and the sea:

> The Sea said
> "Come" to the Brook -
> The Brook said
> "Let me grow" -
> The Sea said   5
> "Then you will
> be a Sea" -
> "I want a Brook -
> Come now" -
> The Sea said   10
> "Go" to the Sea -

> The Sea said
> "I am he
> You cherished" -
> "Learned Waters -   15
> Wisdom is stale
> to me" -

                                        A432/431, F1275B/J1210/M515-516

In a copy sent to Higginson, Dickinson separated the poem into two stanzas between the Sea's "Come now" of line 9 and "The Sea said" of line 10. The space between reinforces a sense of a delay between two sets of discussion. In the first, the Brook is resisting the Sea's order to "Come now." In the second, its implicature is that the brook has refused and grown into a sea which is now ordered to "Go" because it is not what the Sea wanted. The brook/sea then reminds the Sea that it was originally wanted. But the Sea's reply is that what the Brook has learned in growing is not the wisdom the Sea wanted from the Brook as brook.

In the following poem, the poetic self as "I" talks with Christ in alternating stanzas until the last stanza brings them together, in a poetic version of New Testament doctrine:

> "Unto Me"? I do not
> know you -
> Where may be your House?
>
> "I am Jesus - Late
> of Judea -
> Now - of Paradise" -
>
> Wagons - have you - to
> convey me?
> This is far from Thence -
>
> "Arms of Mine - sufficient
> Phaeton -
> Trust Omnipotence" -
>
> I am spotted - "I am
> Pardon" -
> I am small - "The Least
> Is esteemed in Heaven the Chiefest -
> Occupy my + House" - Breast -

                                             H209, F825/J964/M407

Each of the poetic self's questions reveals a voice that is uncertain of the promise of redemption and heaven until the final stanza culminates with Christ's promises in the

Beatitudes of the *Gospel According to Matthew*. Here, the focus is on Christ's promises rather than on Dickinson's own religious uncertainties.

In the following example, opposition is raised to a clichéd pronouncement, in which the first two lines are rebutted by the last two, as the "I" represents Dickinson as poet:

> A word is dead, when it is said
> Some say -
> I say it just begins to live
> That day
>
> Atr66, F278A.2/J1212/M702

Similarly, the following poem, rebutting the cliché that time heals all wounds, opposes the generalized voice of the poetic self against the opening line:

> They say that "Time assuages" -
> Time never did assuage -
> An actual suffering strengthens
> As Sinews do, with Age -
>
> Time is a Test of Trouble -
> But not a Remedy -
> If such it prove, it prove too
> There was no Malady -
>
> H163, F861A/J868/M395

That Dickinson sent the last stanza to Thomas Wentworth Higginson in a letter thanking him for his commiseration over the death of her dog, Carlo, shows how, by putting her last stanza within a life context, she is able to apply the poem's generalization to a personal grief. It is this generalizing characteristic of Dickinson's poems that enables us to respond similarly in recognizing our own experiences in them.

Not all examples of antiphonal play reflect Dickinson's responses to stereotypical pronouncements of her day. Some, like the antiphonal church chants, are rather conversational explorations or elaborations of a particular topic. Others represent an inner dialogue with the poetic self. In the following poem, the three antiphonal questions, "But - what of that?" were underlined in an earlier copy sent to Dickinson's sister-in-law Sue:

> I reason - Earth is short -
> And Anguish - absolute -
> And many hurt -
> But - what of that?
>
> I reason - we could die -
> The best Vitality

Cannot excel Decay -
But - what of that?

I reason, that in Heaven -
Somehow - it will be even -
Some new Equation - given -
But - what of that?

H63; F403B/J301/M215

The first three lines of each stanza reflect the logical reasoning of experienced observation, ending with the reasoned assumption of believers in Heaven that all things will right themselves there. But the fact that each stanza ends with the question, "But - what of that?" foregrounds and emphasizes a new element: what such reasoning means to us emotionally. This antiphonal voice changes its affect as the poem proceeds. At first, it seems with its modern equivalent "So what?" to be impervious to communal suffering; the second to the recognition of inevitability, "What can we do about it?" The poem climaxes as the question changes its affect once more to undermine the comforting reassurance—already destabilized by the "Somehow"—of belief in Heaven: "What good would that do us then?"

But antiphonal play in Dickinson's poetry is much more subtle and pervasive than even these examples suggest. Even when no opposing voice is obviously present, a reader sensitive to antiphonal play can hear an alternative voice.

3 Dramatic Monologue

Whereas antiphonal play in Dickinson's poems involves more than one voice, other poems adopt the perspective of one character in dramatic monologue, much as Robert Browning does in "My Last Duchess." Dickinson will often adopt a male narrator. The speaking persona may be a boy, as in *A narrow fellow in the grass* (F1096/J986/M489). The first version of *The bible is an antique volume* was headed "Diagnosis of the Bible, by a Boy -" and Dickinson sent a longer version, "Sanctuary Privileges" to her nephew Ned (A378, F1577*A*/J1545/M636). In a letter describing Ned, she wrote: "Ned tells us that the Clock purrs and the Kitten ticks. He inherits Uncle Emily's ardor for the lie" (L315).

Although the speaking persona of the following poem has often been understood as a girl, I read it as a boy:

Over the fence -
Strawberries - grow -
Over the fence -
I could climb - if I tried,
I know -
Berries are nice!

> But - if I stained my
> Apron -
> God would certainly scold!
> Oh ⌐ dear ⌐ - I guess if He
> were a Boy -
> He'd - climb - if He could!

H38, F271/J251/M134

Readers have assumed female gender from the reference to "Apron." However, small boys in the nineteenth century also wore aprons to protect their clothing. It is the final lines that confirm for me that Dickinson is adopting a boy persona. As Dickinson once wrote to Higginson, "a pen has so many inflections and a Voice but one" (L470). With the line break after "I guess if He," the inevitable pause causes a greater stress to be placed on "He," thus creating a stress contrast: "If **He** / were a Boy - [like me]." Further evidence comes from a parallel construction over the two stanzas, as follows:

> I could climb - if I tried,
> I know -
> -------------------
> He'd - climb - if He could!

Again, the line break in the first stanza places greater weight on "I know -" that contrasts with "if He could!" God can "scold" but can he "climb"? It is a typical Dickinsonian move, challenging God's capability as opposed to her own.

From these few examples, it can be seen that whereas the selves of a poem's assumed personae do not necessarily represent Dickinson herself, the cognitive process (what I call "minding") that presents the persona is of course Dickinson's. The following example shows how a cognitive reading can catch the authorial attitude behind that of the presented persona.

In response to a contemporary legal case involving the *Springfield Republican*, the First Congregational Church, and the Dickinson family, as well as other members of the Amherst community, Dickinson sent a confidential "In petto -" (in secret) poem to Sue under the guise of Charles D. Lothrop, a local minister accused of domestic abuse:

> In petto -
> A Counterfeit -
> a Plated Person -
> I would not be -
> Whatever Strata
> of Iniquity
> My Nature underlie -
> Truth is good
> Health - and

> Safety, and the
> Sky -
> How meagre, what
> an Exile - is a Lie,
> And Vocal - when
> we die -
> >> Lothrop -

H B157, F1514/J1453/M722

Lothrop sued the *Springfield Republican* for libel, thus diverting the case from the question of abuse to one of language.[1] The irony of the poem lies in the speaker (Lothrop)'s admission that levels of iniquity do indeed "lie" (in both senses) under his nature. A reading of the poem as simply one of Lothrop defending himself is complicated by ambiguity in the poem's structure. Although ostensibly it is Lothrop who is speaking, another antiphonal voice is heard.

I provide a reading for each arrangement that reveals the cognitive dissonance between Lothrop and Dickinson. In its manuscript line breaks, the poem is (1) a reverse sonnet, with the "turn" of the sonnet occurring after line six. This 6–8 line pattern is complicated by (2) an alternative pattern in which the first three and the last four lines create a frame for a central section of seven lines. In reading the poem's structure as a reverse sonnet (1), the first six lines constitute one sentence, which self-describes the speaker:

> A Counterfeit -
> a Plated Person -
> I would not be -
> Whatever Strata
> of Iniquity
> My Nature underlie -

Characteristic of Dickinson's style is putting the object of a sentence first, thus "preposing" it from its normal English order of subject-verb-object (SVO) to an OSV structure. This occurs twice in the first six lines, with the reverting to normal order producing "I would not be a counterfeit: a plated person," and "strata of iniquity underlie my nature." The meaning implied (its "implicature") arising from these statements is "I will therefore tell the truth." But the way Dickinson uses "would not be" instead of "am not," and the way she orders the two sentences and links the two by "Whatever" reveals a more complex admission: I don't want to be seen as someone who conceals the truth, even though my nature leads me into wrongdoing.

In the following section the speaker then elaborates on the first by appealing to the better nature of truth: an entreaty that he in fact adheres to truth. In doing so, he

---

[1]   A full account of the case may be found in Barlow et al. (2014).

is using language to represent himself as someone who sees the goodness of truth, its healing power, its safety, and the promise of heaven (he is, after all, an ordained minister):

> Truth is good
> Health - and
> Safety, and the
> Sky -

The final section then crowns his defense by asserting his recognition of how the practice of lying exiles one from such truth and the promise of heaven. The implicature is why would he do so, as opposed to the *Springfield Republican* which he claims did so; the following stanza commenting on their lie:

> How meagre, what
> an Exile - is a Lie,
> And Vocal - when
> we die -

In reading the alternative structure (2), the central section becomes one sentence, with the first last four lines becoming a comment on the first three:

> A Counterfeit -
> a Plated Person -
> I would not be -
> - - - - - - - - - - - - -
> Whatever Strata
> of Iniquity
> My Nature underlie -
> Truth is good
> Health - and
> Safety, and the
> Sky -
> - - - - - - - - - - - - - -
> How meagre, what
> an Exile - is a Lie,
> And Vocal - when
> we die -

The first three lines may be read the same as in (1):

> A Counterfeit -
> a Plated Person -
> I would not be -

The second three, however, attach to the following four lines of the next section, thus removing the implicature presented in the first reading:

> Whatever Strata
> of Iniquity
> My Nature underlie -
> Truth is good
> Health - and
> Safety, and the
> Sky -

This time, more weight is placed on truth as opposed to the iniquity which may or may not exist, thus Lothrop's attempt to portray himself as a truth-telling, good person. Here, Lothrop's defense rests. As a result, the final four lines stand apart as Dickinson's comment on Lothrop's whole preceding statements:

> How meagre, what
> an Exile - is a Lie,
> And Vocal - when
> we die -

Instead of these last lines being Lothrop's comment on the libel he accused the *Springfield Republican* of, they become an ironic commentary on the previous ten lines attributed to Lothrop. With the first three lines attached to the last four as a frame for the central section, the "would not be" resonates as the poetic self is distinguished from the self-deceit of the speaker.

Even though the Church decided not to excommunicate Lothrop, he nevertheless was exiled from the community. Sue was one of the witnesses deposed in the case and would have understood the deliberate ambiguity of the two readings. It is Dickinson's own voice that emerges at the end. By recognizing the cognitive ambiguity in the two readings, Dickinson can sign her poem "Lothrop -" and comment on his behavior at the same time.

Cognitive readings of Dickinson's poems allow one to distinguish the various presentations of self that are occurring. These can be human or natural beings, the assumption of a particular character or role, a position taken by a poetic self often associating itself with a plural "we" within the context of the situation, and, above all, the attitude of Dickinson herself emerging from these various portrayals.

The Way We Map

One has to learn to read, as one has to learn to see and learn to live.

—Vincent Van Gogh

The way poets think is the way we think. When we write or read a poem, we are using our cognitive faculties. These faculties arise from our sensory relationships with the external world, the visceral and motor activities of our bodies, and our emotive functions, all of which underlie our conceptual reasoning. We are also affected by our experiences, our memories, the cultures and societies in which we live, and our education. Through these cognitive capabilities, mapping across different mental spaces enables us to analogize, perceive identity, formulate our feelings, and so on.

Understanding what it is that human beings do when they make analogies is crucial to understanding what makes us human. Holyoak and Thagard (1995) compare the cognitive capabilities of a chimpanzee and a child.[1] They conclude that although a chimpanzee is capable of certain levels of mapping, a child is capable of much more. Chimpanzees are capable of attribute mapping and—with some extensive training— can do relational mapping, but only the human child can map at the system level.

On the most basic level, attribute mapping occurs when we recognize the sameness of objects, like identifying the student who walks into the office one day as being the same student who came in the day before, or recognizing Dickinson's reference to "Rose" in the following line: "If I should cease to bring a Rose" as being the same as in the line "'Twill be because beyond the Rose" (A80-3/8, F53/J56/M45). This basic level extends to recognizing that two different names can refer to the same individual, as when Dickinson refers to a "Mouse" in one poem and then calls it a "Rat" three lines later (H7, F151/J61/M90). Attribute mapping also occurs when we recognize the

---

[1] Holyoak and Thagard's study had the ancillary effect of shedding new light on the argument between linguists and psychologists as to whether chimpanzees had been able to break through the language barrier that Chomsky and other linguists claimed differentiated the human species from its closest relatives. Holyoak and Thagard chose to study the findings of David Premack's research with his chimpanzee Sarah, since she had been brought up in laboratory conditions and taught to deal with functional analogy problems (Premack and Woodruff 1978). The Gardners' chimpanzee, Washoe, on the other hand, had been raised in a domesticated environment and taught sign language (Gardner et al. 1989). Most of the dispute between psychologists and linguists centered on the more ambiguous question of whether Washoe and other chimpanzees like her had actually mastered the rudiments of language. The data available from the experiments with Sarah more directly reveal the limits of a chimpanzee's natural cognitive capacities.

similarity of objects, such as identifying several pieces of fruit in a basket as apples, or two animals running down the road as coyotes.

With relational mapping, analogy operates at a more abstract level, with relations between—rather than the objects themselves—being highlighted or "profiled." It depends on the ability to distinguish the differences between objects as well as their similarities, and yet to perceive that they can have relations in common. In Dickinson's Snowstorm poem, for instance, the falling of snow on trees is likened to the action of sifted flour powdering a board, and the gray, snow-laden clouds to lead-colored, heavy sieves: "It sifts from Leaden Sieves - / It powders all the Wood" (H278, F291/J311/M248). In another poem, the actions of two people who die for beauty and truth, respectively, cause them to become related as "Brethren" in the tomb (H183, F448/J449/M225). When Dickinson calls her dog "the noblest work of Art," she is invoking the relationship of comparison. Relational mapping is the basis for metonymy and metaphor, as in Dickinson's use of "Sandals" for the bottom (foot) and "Bonnets" for the top (head) of the mountains (A83-7/8, F108/J124/M77).

Once multiple mappings begin to occur, the relations between the components of the mappings become more abstract so that it is the structure of the relations that is seen to be the same or be "isomorphic" (having identical form), and mapping occurs at the system level, drawing on the knowledge domains of Idealized Cognitive Cultural Models (ICCM). However, as Holyoak and Thagard (1995, 100) note, though "adults have the potential for creative mental leaps that break the bounds of simple similarity, yet they also continue to be guided—and sometimes beguiled—by surface resemblances."

1   Historical Background

When I was teaching at Los Angeles Valley College, the Math and English Departments shared a lab for computer-assisted instruction. This close proximity led to a collaborative project in which we tried to discover why students were experiencing difficulty with word problems in beginning Algebra. Since so many students were non-native speakers of English, the Math faculty thought it might be a language problem.

I quickly learned that it made no difference how many times I changed the wording of the question, the students still had problems. I suspected that the trouble was not so much the result of the language in which the problems were couched but in the way the students were applying the cognitive skills needed to solve them. In order to arrive at a better understanding of how human cognitive skills were being applied and based on a test that the chimpanzee Sarah was never able to solve, I designed a quiz for the students (Figure 11.1). The only difference between the two was that I could ask the students why they chose one alternative over the other, something that of course Sarah could not do.

If one applies attribute and relational mappings to the question of which alternative better matches the sample, the tendency will be to choose alternative B, which at

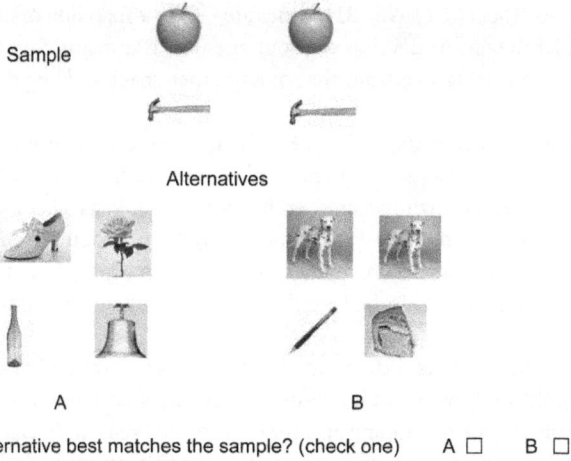

Figure 11.1 System mapping test (based on Holyoak and Thagard 1995, 70).

least shows a 50 percent match between the upper pairs of objects. However, a more complex and sophisticated mapping, based on isomorphic (same structural) relations of relations, would result in choosing alternative A. That is, the relation between the upper pair of objects in the sample to the upper pair of objects in A is same-different, and the relation between the lower pairs is also same-different. Therefore, based on a system mapping interpretation, alternative A would be the better choice (Figure 11.2).

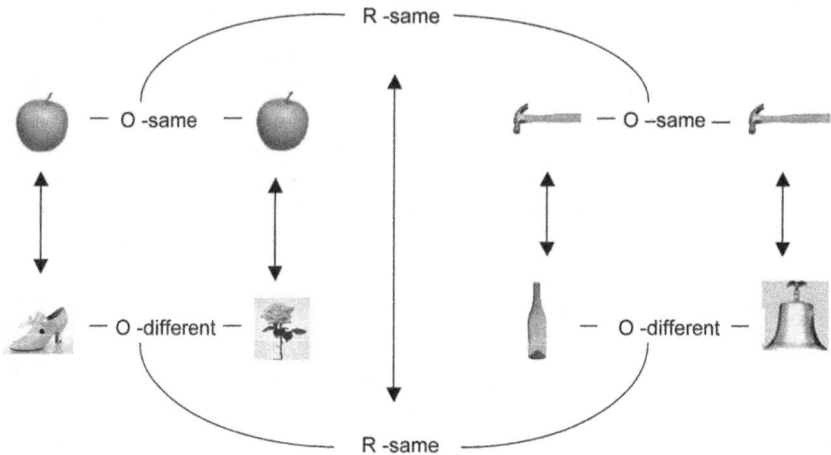

Figure 11.2 System answer: A (based on Holyoak and Thagard 1995, 71).

Holyoak and Thagard (1995, 81) conclude: "The capability of finding system mappings, which begins to develop at about age five, is a major cognitive transition separating human intelligence from that of any other species." How did the students perform?

The students overwhelmingly chose B. When a few did choose A, their reasons were similar to the reasons given for choosing B. That is, they all mentioned features attached to the objects shown (organic, edible, tool, etc.). What this told me was that our students were not spontaneously applying system mapping; rather, they were applying attribute and relational mappings to the problem, something that Sarah, after a great deal of training, was also able to do. Were the students no better than trained chimpanzees?

Intrigued by these results, I distributed the test to as many staff and faculty at the college as I could find, with similar results. Only one staff member—our computer specialist—applied system mapping and chose A for that reason. In case you run away with the idea that people at my college were unusually cognitively deficient, I should note that I then took my quiz on the road, to graduate students, local teachers, and faculty at Lancaster University in England, to an International Cognitive Linguistics Conference in Amsterdam, and to faculty, staff, and students in a Humanities seminar at Colby-Sawyer College in New Hampshire, with similar results. Not being a social scientist, I did not have the expertise to set up the necessary controls and do statistical analysis that would normally be demanded for a formal experiment. However, I believe my informal survey demonstrates that people on the whole spontaneously use attribute and relational rather than system mapping in approaching an analogy problem like the one given to Sarah. Of course, when presented with the system answer, everyone was able to see the solution.

Although my findings have obvious pedagogical implications for education, my own research interests led me to apply my findings to the ways in which readers adopt mapping strategies in understanding and interpreting poetry. What I have discovered is that differing interpretations of poetry depend on both the kind and level of mapping strategies adopted (M. Freeman 2002a). System mapping, as the experiment shows, demands the more abstract cognitive capability of seeing structural, isomorphic relations between different objects and concepts. Although the Sarah experiment deals with visual objects, system mapping also applies to analogies used in natural language to create new meanings. Readers construct new meaning and significance from a poetic text through mapping across mental spaces (Fauconnier and Turner 2002). Blending is a dynamic cognitive mapping process that models the way we project from several mental spaces to a new blended space that has emergent structure not present in the original spaces (see Chapter 9).

As Fauconnier and Turner (2002, 158) note, although interpretations of a language expression may differ, basic mapping strategies used are the same. The actual mappings might differ as a result of selecting different counterparts for the cross-space mapping and performing different integrations in the blend (168). These choices partly depend on the use of background knowledge, context, and long-term memory, but little work has been done to date on how these and other factors might constrain interpretation.

A still relatively unexplored question is what, in any given interpretation, determines the choice of input space domains and what gets selected in them for projection to the blend. A closer examination of the mappings used by readers of poetry would thus explain not only how readers are able to find different meanings in literary texts, but also how they construct the blends that yield these meanings (Freeman and McLoughlin 2021). As my discussion of the Sarah experiment shows, different interpretations arise not simply from the use of knowledge, context, and memory, but also from application of the three mapping strategies of similarity, structure, and purpose.

2 Reading Readers Reading a Poem

A brief survey of literary interpretations of a Dickinson poem reveals that readers in fact use some, if not all, of these mapping strategies. For my data, I selected a sample of discussions of an Emily Dickinson poem that were posted to the emweb listserv in May of 1999.[2] Marcela Linkova, a graduate student from the Czech Republic, raised questions about the poem, "Of God we ask," on page 150 (Figure 11.3).

Four of the six posts generated by this query provide direct interpretations; the other two give contextual evidence from other Dickinson writings. The somewhat different interpretations arise from slightly different conceptual and metaphorical mappings. Linkova writes:

> It seems to me that the argument is pretty straightforward. She criticizes the original sin and our required asking for forgiveness to a "crime" that is never really revealed to us. She seems to question the image of this life as magic "prison" where we reproach happiness in view of the bliss to come. It is, however, the "too" in the last line that I find enigmatic. I was thinking that there is a transposition of the too, meaning "as well," suggesting that happiness competes with Heaven just like we, perhaps, compete with God. Another alternative, even more enigmatic, is that there is an elision of "much" after too, and the line should sound "That too much competes with Heaven"—but WHY would happiness be competing too much and how that would be done? Could it be that because the happiness tries too hard, it is bound to lose to Heaven?

Linkova maps the phrase "a magic Prison" onto life, "Crime" onto original sin, and "Heaven" onto bliss (which is greater than happiness). She sees the stance of the speaker as one of protest against these mappings. She raises the question of whether the word "too" in the last line should be read as meaning "as well" or "too much." Reading "as well" for "too," she produces a double analogy that just as happiness competes with heaven, we compete with God. Her comments suggest that she understands our

---

[2] Posted 01:03:55 EDT 97-05-17 by Ruth Gustafson to EmMail1@aol.com (no longer in existence). I have silently corrected typographical errors. All are cited with permission.

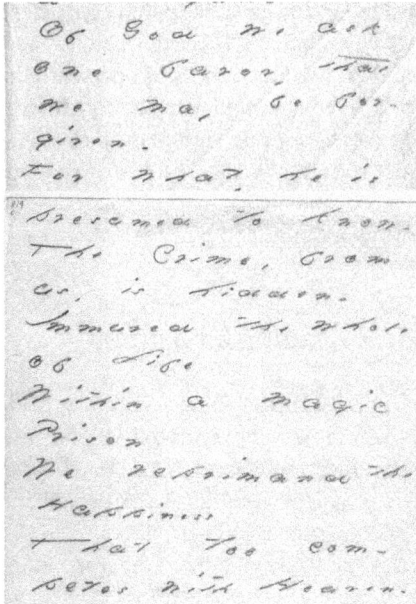

Of God we ask
One favor, that
we may be for-
given -
For what, he is
presumed to know -
The Crime, from
us, is hidden -
Immured the whole
of Life
Within a magic
Prison
We reprimand the
Happiness
That too com-
					petes with Heaven.

Figure 11.3 *Of God we ask one favor* (A819, F1675B/J1601/M655) Manuscript Courtesy of Amherst College, Emily Dickinson Collection.

competition with God as one over whether his mappings (of life as a prison for the crime of original sin) should take precedence over ours (which is presumably that the goal of life is happiness). She uses the same mappings when she reads "too" to mean "too much." This causes her to ask why happiness should be trying so hard and whether this means "it is bound to lose to Heaven."

The second post, by Wayne Whittaker, a high-school teacher from the United States, adopts the "too much" scenario and points out that Linkova's composition of the comparison is incomplete: "I would say your second interpretation is closer to the mark, and that the meaning is that terrestrial happiness by competing for our attention causes us to turn from the pursuit of eternal happiness in heaven." Competition, as Whittaker notes, does not exist in a vacuum but is competition "for" something. He maps that something onto "our attention" to produce a resolution for Linkova's question. In doing so, he changes Linkova's cross-space identification mappings of *happiness = us / heaven = God* to a PURSUIT metaphor involving a PATH schema in which life deflects us from trying to catch up with heaven.

In Whittaker's reading, "magic" is understood as something alluring and something to be desired. Bonnie Poon, a high-school graduate and entering college freshman from Hong Kong at the time of her post, understands magic as ritual that we need to avoid:

Maybe the "magic Prison" refers to the religion which can play such a large role in forming the society. Perhaps ED sees it as "magic" because she questions the religious part of religion—i.e., making a ritual out of a relationship with God (?—I'm not an expert on ED's religious standing) which can so control people and has a sort of attraction of its own. And because religion teaches people to seek not what is of this world (the "Happiness"), we have to reprimand, rebuke what earthly things that bring that Happiness which is not of God.

I think both readings could work, though one might be a little more farfetched than the other. The one that suggested "too much competes with Heaven" is the way earthly desires block one's relationship with God. Perhaps she is speaking of the way it is difficult for humans to "let go" and just go full on for God, and so God is, essentially, competing with the world for humans.

The reading which suggested "also" competes with heaven: if we substitute "too" with "also," that in itself suggests that there are two things that compete with heaven—one being Happiness, and the second might be Prison (it's capitalized too), which goes back to the whole religion being too religious and not personal, thus becoming a Prison rather than a gate to freedom.

Poon attempts to accommodate the readings resulting from the different construals of "too" by creating another set of mappings. She maps "magic Prison" onto religion by metonymically relating "magic" to "ritual" that can control and attract and maps "Happiness" onto worldly things or "earthly desires." Both these mappings (religion and earthly desires) compete with heaven, which she metonymically identifies with God. Under these readings, God competes for humans with both world happiness and the ritual of religion. Poon draws on an Idealistic Cognitive Cultural Model in her understanding of Christian religion—that Christians are taught to "seek not what is of this world" and that religion is the "gate to freedom"—to distinguish between that and false (i.e., ritualistic) religion that is "too religious" and therefore becomes instead a "Prison," which blocks our relationship with God.

The fourth post, by Stephen Browning, a classical musician from the United States, questions Linkova's mapping of "Crime" onto original sin by introducing biographical details:

I'm not sure that I can agree with your "original sin" notion. Did ED ever embrace the concept of original sin? It seems very far removed indeed from her daily wrestling with the idea of God, and her creation of him in so many different guises. Actually, the notion of original sin strikes me as exactly the kind of conventional dogma that she so strongly rebelled against in her teens, as abundantly documented in the early letters.

I think that there is a paradox operative in this poem—on the surface it would appear that the "Crime" is that we love life too much, and that that love distracts us from higher matters. But then the poem appears to contradict itself, since it asserts "The Crime, from us, is hidden." One can't really have it both ways. Underlying the paradox, I think, is the notion that God himself has created this situation for us, it is he who has created this "magic Prison" of life, has made it so alluring that it bids to outshine heaven, and we experience guilt on this account. But this is

GOD's doing, not ours, and ED knows it. So she goes on asking for forgiveness, knowing that she can't know what it is that must be forgiven. The circularity of ED's reasoning is characteristically brilliant, and is a trap from which there is no escape.

Of course ED wouldn't be ED if there were only one possible interpretation. On another level the poem is certainly a heartfelt expression of the need to beg forgiveness for enjoying life too much (her "paganism").

Browning introduces the cultural knowledge domain of Dickinson's religious background and her self-identification with "paganism," and offers a more complex reading by mapping the "Crime" onto "lov[ing] life too much." This produces a paradox, since the crime is said to be unknown to us, but then the last lines say what it is. To resolve the paradox, Browning maps God as Creator of the prison, which is a metaphor for life, and makes him responsible for the "Crime" of creating a world that can "outshine heaven." His conclusion, that the circularity in reasoning produced by the paradox is "a trap from which there is no escape," results from understanding the poem as a complex double-scope blend in which topology is being projected from both input spaces.

In one input space for the metaphor of LIFE IS A MAGIC PRISON, there is the prison itself, with its associated components of a prisoner who has committed a crime and who (in Browning's reading) experiences guilt and asks forgiveness. In another space, life is understood as being "alluring," causing us to love it too much. The mappings across the spaces link the prisoner to us, the creator of the prison to God, and the prison itself to the world in which we live, as follows:

crime	—— human behavior
plea	—— forgiveness
judge	—— God
accused	—— us
sentence	—— life on earth

The metaphor produces a blended space, expressed in the phrase "magic Prison." This space produces emergent structure that occurs in neither of the input spaces. That is, prisons are not alluring; being born is not a crime (Browning rejects Linkova's notion of original sin); enjoying life does not produce guilt; we can't ask for forgiveness for something we haven't done; and we can't feel guilt for a crime we didn't commit. The emergent structure that makes all these things happen enables us to understand life as a world, unlike a prison, that we do not want to leave, but because we love it so much, we have to ask forgiveness for not wanting heaven instead. In the following diagram, identity mapping links items in the incarceration space with those in the life space (Figure 11.4).

In the blend, because we are asking for forgiveness, God is represented as both victim and judge. Browning's paradox produces another blend in his interpretation: we want to and can escape from prison, but we can only escape from life through death,

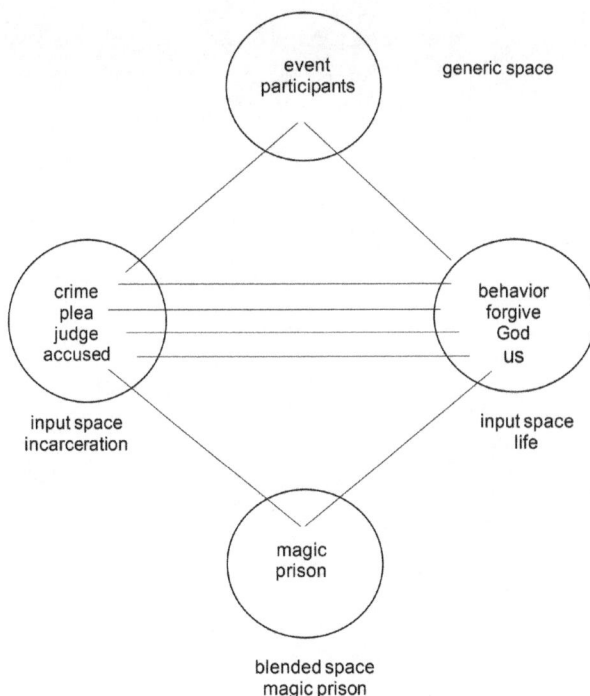

Figure 11.4 The complex blend in *Of God we ask one favor* (A819, F1675B/J1601/ M655).

which we don't want. The life God has created for us is "a trap from which there is no escape." That is, in the poem's "Prison," we are given a "life" sentence.

This preliminary rough sketch of these postings suggests that cognitive analysis can provide a means whereby literary interpretations might be described and categorized with respect to how cultural knowledge may influence readings and which readings might be more prototypical with respect to a poet's life and other writings. The interpretations themselves provide a database for empirical research on how metaphorical meanings are mapped and what motivates them. Cognitive analysis shows that the readers are using metaphorical and metonymical mapping strategies and their own cultural knowledge and backgrounds to understand the poem. Such studies reveal the kind of mental operations that contribute to skillful analysis and interpretation.

By linking literary analysis with the general processes of human minding, a cognitive approach to literature shows how human creativity can be nourished by reading poetry. As Browning says in his second post, Emily Dickinson "trains us to think in metaphors and make fantastic leaps of logic, so almost anything COULD appear to be connected to anything else." Blending, in particular, shows exactly how important metaphorical thinking and Browning's "fantastic leaps of logic" are to human creativity.

Intentional Mapping

The Search for Coherence

One universal trait in all human cognition is the search for coherence. As a result, we tend to leap to interpretation, without always considering how much we are affected by our subliminal motivations. As Hannah Fry (2021, 71) puts it, "Our perspectives are hard-coded into what we consider worth considering." As we construct meaning from a poem, we too often overlook the ways our own biases and perspectives are active in formulating our reading. In other words, the way we map is partially determined by the way we are predisposed to think a certain way. As a result, we do not always consider our biases in the way we interpret a poem or in the way we focus on what may have been the poet's subliminal intensions.[1]

For someone who loved life as much as Dickinson did, it can seem puzzling to readers that the word *death* occurs so high on the list of word frequency in her poems (Rosenbaum 1964). It is indeed the subject of many of her poems and has elicited much critical analysis. There are three major interpretive approaches to Dickinson's views of death that suggest different and conflicting statements of belief: there is life after death; there is no life after death; we can never know whether there is or isn't. Is Dickinson, then, an agnostic, an atheist, or a believer? Critics have argued for any one of these positions. In doing so, they are looking at what the poems mean rather than what the poems do, a mistake that treats expressive, emotive language as though it were communicative and discursive. By missing the point of artistic purpose, critics end up creating their own meanings as though they were the poet's.

One poem that has drawn a great deal of attention is *I heard a fly buzz when I died* (A84-1/2, F591/J465/M270). Examination of some characteristic readings of this poem reveals that the mapping strategies critics are using to interpret it depend on their beliefs about death.[2] My initial hypothesis—that a blending analysis might lead

---

[1] The term *intension* originally meant a stretching and straining related to the voice in music. It is related in logic to the *internal* quantity or quality of a notion or concept. The *OED* cites William Stanley Jevons (1876): "In putting steam before ship, we have greatly reduced the extension of the term. But we have increased its intension, because steam-ship means *all that the ship does, and more, for it means that the ship is moved by steam power*" (v§23.22; my emphasis). See also note 3.

[2] My data is drawn from two sources: Joseph Duchac's (1979) annotated guide to Dickinson's poetry, which contains sixty-one entries on the poem from 1949 to 1977, and a selected array of criticism

to the possibility of evaluating readers' insights about Dickinson's own attitude toward death—proved to be wrong. There was no independent way based on conceptual blending alone to determine validity among the various interpretations I examined. I realized that a closer study of the poem's prosodic and affective characteristics was needed.

1 Three Case Studies

Dickinson's Fly poem presents a typical nineteenth-century deathbed scene, in which family and friends gather in the death chamber to witness the passing of a dying person (Hogue 1961). Its peculiarity arises from the fact that the scene is presented from the viewpoint of the person who is dying:

> I heard a Fly buzz - when
> I died -
> The Stillness in the Room
> Was like the Stillness in the Air -
> Between the Heaves of Storm -
>
> The Eyes around - had wrung them
> dry -
> And Breaths were gathering firm
> For that last Onset - when the King
> Be witnessed in the Room -
>
> I willed my Keepsakes - Signed away
> What portion of me be
> Assignable - and then it was
> There interposed a Fly -
>
> With Blue - uncertain - stumbling Buzz -
> Between the light - and me -
> And then the Windows failed - and then
> I could not see to see -

A84–1/2, F591/J465/M270

The differing interpretations critics have given to Dickinson's poem all depend on the selection and suppression of various metaphorical schemas on all three levels

from 1977 to the present time. I sorted the various analyses of the poem into groups according to their metaphorical or symbolic interpretations, and examined each of these groups to determine whether the readers were using the same mapping paths to arrive at their conclusions. I then compared these conceptual mappings across the various groupings.

of image, structure, and purpose. The reasons and explanations the critics give for their selection of equivalences for the fly, the King, the dying persona, and so on, all involve constructing scenarios in which the language of the poem is made to achieve some kind of interpretive coherence. These scenarios often involve the development of structural analogy. At the level of mapping by identity, critics are almost equally divided between those who see Dickinson's fly as a symbol of life and those who see it as a symbol of death. Some see it as both. Likewise, whereas some equate the fly with Satan or evil, others identify it variously as the soul, as the "King" of stanza 2, as putrefaction, as nature in its most trivial aspect. These equations are all operating on the level of attribute mapping (see Chapter 11).

The reasons critics give for creating similarity between the fly and the soul depend on their construction of an analogy between the fly in the death chamber and the soul in the body. These critics are adopting a CONTAINMENT schema in order to create an analogous relationship: just as the fly buzzing at the window is attempting to escape the confines of the room, so the soul at death is attempting to escape the confines of the body. Note that in the poem there is no direct evidence that the fly is actually at the window; this scenario has been introduced by the critics from experience: memories of seeing and hearing flies at windows in rooms. Nor is the analogy necessarily perfectly isomorphic. Some critics make a logical inference that the meaning of the poem lies in the denial of an afterlife, since as the fly cannot escape (presuming the window to be closed) neither can the soul. Others who make the fly-soul identity connection nevertheless argue for the poem asserting belief in an afterlife.

The interpretation given to a poem is itself the construction and elaboration of meaning through the operation of double-scope blending, with projections made from the various input spaces set up by the reader. These projections are constrained by the intention or motivation of the particular critic.[3] Invariably, these interpretations conflict, not just at the level of attribute or relational mapping, but at the system level mapping of purpose.[4] In the following sections, I explore three conflicting interpretations in order to determine to what extent mapping strategies, constraints on identification of relevant spaces, and selection of topology for projection are guiding the critics' readings.

Case 1: The Portal to Nowhere: Death as Disintegration

Several critics take the position that the poet's communicative intention in the Fly poem is to assert a nihilism of total extinction. Brooks, Lewis, and Warren (1973,

---

[3] The terms *intention* and *motivation* are sometimes conflated and used interchangeably, although I see them as two distinct cognitive processes. As R. W. Gibbs (1999, 80) notes, "An intention is a mental representation caused by a desire for a goal which itself causes action to bring about that goal." Intention thus refers to the "goal" or meaning—the "what"—the writer or reader wishes to communicate; motivation to the "desire" or emotion—the "why"—that invests meaning with significance. Motivation in this sense is therefore more akin to subliminal "intension" than conscious intention (see note 1).

[4] R. W. Gibbs (1999, 234–72) provides a succinct summary of the different theoretical assumptions that underlie the various literary critical approaches to the question of authorial intention.

1249–50) perhaps express the point most bluntly in their claim that the poem reveals "a disconcerting truth: that one's death may be a most trivial event, hedged about with irrelevancies, and leading to no afterlife." In these readings, metaphors of light and seeing as understanding constrain the ways in which the fly is perceived. A focus on perception leads readers to select the poem's images that most directly link to sight: the eyes, the synesthetic use of color in the "blue buzz" of the fly, the light, the windows (G. Johnson 1985, 166). Wardrop (1996, 192: n. 1) claims that Dickinson's use of the term *window* in her poetry "nearly always makes an epistemological statement" in that "the window bifurcates the world, aligns perception into terms of inner and outer, posits a necessary duality constantly policed by the beholder." For Wardrop, the poem's last line "locates a kind of epistemological nihilism."

Such readings invoke the conventional metaphors that link light to seeing and seeing to knowing, so that failure of light means failure of life. The intrusion of the fly becomes a relational one. Kher (1974, 206) notes:

> Death seems to mean here the total loss of perception: the windows fail and the persona cannot see to see. The last conscious link with reality is established through the buzz of the fly. The humming of the fly or song of death unsettles the relationship between the dying person and the light. In death, then, the dying person does not experience anything at all concerning life after death. The blue, which is Dickinson's symbol for eternity, becomes here the symbol of complete extinction.

It seems that critics who take the position that the poem is a statement about total extinction are applying both attribute and relational mappings in elaborating the poem's metaphoric blend. The fly in these readings blocks the light of celestial revelation and therefore stands for final extinction.

Case 2: The Portal to Immortality

Many critics have situated Dickinson's Fly poem within the cultural conventions of nineteenth-century belief in the significance of the deathbed scene to prefigure the soul's salvation (Anderson 1960; G. Johnson 1985; Olney 1993; Hogue 1961, among many others). Sharon Cameron (1979, 115) reverses the meaning other critics have constructed from the poem's last line, "I could not see to see -," by claiming the central premise of the poem as "death is survived by perception." She constructs a divided persona: one who has a false preconception of death that "gives way only to darkness" (114) and a "more knowledgeable" one who finally comes to see the reality of death in surviving it (208). The fly becomes the enabler ("the King") who obscures life, allowing the speaker in not seeing in death to see beyond it.

Unlike Cameron, Emily Budick (1985, 187) attributes an alternative and opposing view, not to a divided persona, but to the poet herself:

> Thus, in "I heard a Fly buzz," it is not the poet of the poem who accepts the analogical premises of the "Heaves of Storm" philosophy and who is therefore

carried at the end to perpetual eclipse. Rather, it is the narrator who errs in her symbolic reasoning. For the poet and for the reader the fly can communicate an authentic vision, and it succeeds in communicating this vision symbolically: like all symbols, the fly brings the noumenal and the phenomenal into some kind of relationship.

In other words, the disanalogies of counterfactual mappings Budick discerns are not between the conflicting symbolic interpretations attributed by a divided persona to the elements of the natural world, but between "the earthly and the transcendent" (188): "In being made to emphasize the difference between itself and God, the fly is able to confirm for us the existence of the 'King,' the same king whose existence is eclipsed by the same fly when its symbolic meaning is misinterpreted by the narrator of the poem" (187–8). Or, one might say, by its readers.

Case 3: The Limits of Experience

The fact that we can never know for sure what, if anything, happens to us after we die underlies the last set of readings. However much we try to strive after the unknowable, however much we might desire to believe in an afterlife, such attempts and desires are futile. In the "limits of experience" readings, a forced incompatibility between the two spaces—experience and belief—creates a self-contradictory blend which falsifies the belief space.

In his discussion of death as "experience of the end of experience," James P. Carse (2008, 103–6) sets up the *belief space* of the deathbed scene as one of expectancy on the part of the dying person and the watchers, the "last onset" when "the King be witnessed in the room," typifying anticipation of the soul's release into immortality. The *experience space* contains the physical description of the mourners who appear lifeless—"Those attending are silent and unmoving; their tears have dried; even their breathing has ceased"—as opposed to the fly—"The only obvious presence of life is the fly, wandering the room without knowing why or where ('uncertain' and 'stumbling')." Both spaces share topology: a dying person, watchers by the deathbed, an air of expectancy toward the moment of death when something "solemn and climactic" will happen.

There are also, however, disanalogies which cause conflict in the blend and make one space counterfactual to the other. In the conventional belief space, it is the watchers who are witnessing the passing of the dying person, and thus the scene is enacted from their point of view. In the experience space, the point of view is coming from the dying person, and it is the watchers who appear lifeless. In the belief space, the watchers anticipate the arrival of God or Christ as "the King" to take the dying person's soul. In the experience space, only a fly intrudes. As the watchers live in the blend of the belief space, they project their experience of an actual dying into a blended space by their belief in an *immortality space*, in which God will rescue the soul from its mortal end to a new beginning in Christ, so that the "end" is the "beginning" as the moment of death becomes that "last onset." But since the poem is told from the perspective of the dying persona, the belief space becomes counterfactual, so that it is the experience space that is actual, and there is no projection of a new beginning from the end, nor a God-King arriving to rescue the soul.

In this reading, the focal image of the fly triggers the disanalogies that make the belief space counterfactual. "Flies," Carse tells us, "are to be killed and brushed away." In Carse's reading, there is no "enduring substance," either for the fly or the soul. In the end, both will be brushed away and forgotten. As Carse runs and elaborates the blend that is the poem, his perspective comes from the world of the living who both forget the dead and lose their own sense of community as a culture, so that the conventional religious perspective betrays the true religious perspective of memory and "communitas."

Whereas Carse's reading ultimately focuses on the limited perspective of the living, Jane Eberwein's (1985) reading focuses on the loss of that perspective by the dying. The difference is one of purpose. Whereas Carse uses Dickinson's poem to shed light on "an understanding of our own death," Eberwein's intention in her book on Dickinson is to shed light on "the mind that produced" the poems (ix). For Eberwein, then, Dickinson's Fly poem highlights the limitations imposed upon the human condition: "Imagination—although fused in a blending of dream and drama—could never fully place her [the poet] outside the circuit [of the life-world], however, she might pivot on its brink" (219). Eberwein's blend of "dream and drama" sets up two input spaces. In one, you have the poet attempting to explore the "mysteries" that lie beyond the life-world. In the other, you have a dead person who, being beyond the life-world, has access to those mysteries.

Identity mapping between the two spaces adds value to role, so that the poet plays "the role of the dead person" in order to probe the mysteries blocked for the living poet. However, the disanalogy between life and death creates a conflict in the blend. The fly, in Eberwein's reading, is mapped not onto the soul but onto the "trivial but nonetheless firm clutch of the circuit [life] world" that "blocks celestial revelation" so that it is impossible for Dickinson to "achieve perspective on the mysteries she wanted to probe" (219). Robert Weisbuch (1975, 99) makes a similar point in discussing the adopted perspective of the poem's persona:

> [T]here is no ostensible suggestion that the poem is describing anything "earlier" than how the mind feels when it dies. Yet the scene of this "last onset" is so exclusively the mind that it cannot help suggesting figural, experiential states of consciousness. Thus Dickinson insists that the posthumous voice can remain to speak with us, though logically it should be gone and silent. Her justification is that the voice, by hyperbole, speaks to our living condition.

The three case studies reveal different selection and integration strategies: the "portal to nowhere" readings adopt the KNOWING IS SEEING metaphor to select light-perception and consciousness; the "portal to immortality" readings divide the points of view of persona or persona and poet to invert conventional symbol-making that leads to false preconceptions in order to assert the true symbolism of time and eternity; the "limits of experience" readings select belief and the desire for knowledge to shape topology. All these critics are motivated by their need to provide the poem with interpretive coherence based on their assumptions of the poet's intentions.

What is missing, ironically, in the literary critical analyses of Dickinson's Fly poem is recognition of the experiential dimensions of reading the poem.[5] Feelings arise from three sources:

1. sensations, from interaction with the external world through the five senses;
2. bodily motion, including internal, visceral responses (pain, discomfort, etc.); and
3. internally generated emotive forces that contribute to and arise from the sensory and the motor.

In the following section, I explore the way prosodic affects set the poem's tone and thus lead to a possible way to determine what, at least in this poem, Dickinson is doing with the idea of dying.

2 Tracing the Forms of Feeling

First is the question of motivation: why Dickinson wrote this poem. That of course is an impossible question to answer, but there are clues as to what might have prompted her poem. Samuel Taylor Coleridge once remarked that poets draw from two sources: their experience of nature and their encounter with books. There is evidence that Dickinson played with creating poetic responses to her reading.[6] Hyatt Waggoner (1968, 673–4: n. 4) noticed the echoes of image and language that exist between Dickinson's Fly poem and Hawthorne's (1922[1851]) chapter called "Governor Pyncheon" in *The House of the Seven Gables*. It is as though Dickinson, on reading Hawthorne's description of the Judge's death, thought, "What can I do with this theme? What is it like to experience the moment of death?" A comparison of similarities and differences reveals that Dickinson's mappings of the structures and forms of Hawthorne's chapter illuminate the affective nature of Dickinson's poem.

---

5   By "experiential dimensions," I am not referring to the emotive responses of readers, as studied by many emotion researchers (see Miall 2006 for a sample bibliography), which is a legitimate subject for study in its own right. Rather, what I am interested in is those aspects of the text that serve to trigger those emotive responses, the "formulated feeling," to use Langer's (1953) term, which characterizes the literariness of a given text. Miall (2006, 169) rightly argues for the role of feeling in imagination and shows how Coleridge's theory of the imagination is superior to Kant's in negating the dichotomy between body and mind.

6   It is known that Dickinson more than once tried her hand at constructing her own expressive version of a theme she had met in her readings. For instance, on the occasion of the third anniversary of her father's death, following her comment in a letter to Thomas Wentworth Higginson that "I was rereading your 'Decoration.' You may have forgotten it," she appends a poem which Higginson said later in a letter to Mabel Loomis Todd was "the condensed essence" of his poem and "so far finer" (Todd and Bingham 1945, 130; see Chapter 15).

Hawthorne's conceit in the chapter is to imagine an observer who delays his experiencing recognition of death, even though his descriptions make it clear to the reader that the Judge is indeed dead. Hawthorne thus sets up an ironic situation in which, as the observer lists at length all the things the Judge should be up and doing instead of sitting motionless in his chair, the reader (at least this one) becomes exasperated at the observer's failure to see the plain truth. As the ostensible purpose of this conceit is to allow Hawthorne to focus on the Judge's worldly life, his character and future expectations, it draws the reader into an emotional engagement with the scene: "You must hold your own breath, to satisfy yourself whether he breathes at all. It is quite inaudible. You hear the ticking of his watch; his breath you do not hear" (301).[7] The ticking watch is the only other sound apart from the storm gusts outside and it is "a fearful one" which "has an effect of terror" (311). What terror there is must be experienced by the observer, since the Judge is past all feeling.

What does Dickinson do with the image schemas and structural relationships of Hawthorne's chapter?

First, she fuses the dying person with the observer, so that Hawthorne's narrator, who is observing a person already dead, becomes, in Dickinson's telling, a person in the process of dying.

Second, she maps Hawthorne's address to "you," which engages the reader in the process of witnessing the dead body, onto the watchers by the deathbed.

Third, she interiorizes the scene within the confines of the death chamber so that the storm that is physically raging outside in Hawthorne's chapter is rendered as an analogy to the stillness of the air within the death chamber, so that the "heaves of storm" may more directly be read metaphorically as interior ones.

Fourth, she compresses Hawthorne's "fainter and fainter grows the light" into the windows fading (267).

Fifth, she converts two of Hawthorne's images—the Grimalkin devil cat and the window entry into consciousness and the soul—into the fly interposing itself "between the light - and me -." Such compression inherits the fortuitous accidental associations of a fly as feeding on death, feelings of disgust, and the idea of evil (in the naming of Beelzebub as "lord of the flies").

Finally, she collapses the ticking of the Judge's watch, which is the only sound within the silence of the room, onto Hawthorne's fly on the Judge's dead body to create the incessant buzz of Dickinson's fly. These compressions all serve to shift the point of view from the external observer to the dying person. The effect of this shift is to remove all traces of feeling from the death scene.

As Waggoner notes, Hawthorne's chapter ends on a much more optimistic note than Dickinson's poem, with the light of morning streaming through the windows into the Judge's death chamber. Because the observer in Dickinson's poem has become fused with the dying person, her poem ends at the point that Waggoner (1968, 674) describes as Hawthorne's "darkest mid-point of the chapter":

---

[7] The page numbers in this section all refer to Hawthorne's chapter, except where noted.

There is no window! There is no face! An infinite, inscrutable blackness has annihilated sight! Where is our universe? All crumbled away from us; and we, adrift in chaos, may hearken to the gusts of the homeless wind, that go sighing and murmuring about, in quest of what was once a world! (311)

It is noticeable that the emotions here associated with Hawthorne's living observer are absent from Dickinson's persona's simple, flat statement that "the Windows failed," as is the hint of sensations of putrefaction and decay in Hawthorne's living fly that "has smelt out Governor Pyncheon" (318).

Dickinson's poem creates an ironic inversion in order to characterize the loss of human feeling in death. How is this inversion effected? Ceding, cessation, and stasis structure the poem's form. The poem moves from the dying persona's sensation of sound ("I heard") to the failure of the sensation of sight ("I could not see"), a framing device that serves to contain the deathbed scene within the opening and closing lines as it is contained within the confines of the room. The poem, unusual for the lyric, is couched entirely in the past tense, a movement that distances the experiential effect of the present to the narrative mode of the past, so that any human feeling that might be experienced is already gone. This distancing is reinforced by the introduction of the idea of feelings in the allusions to commonplace formulated expressions (heaving sighs of grief, wringing one's hands in grief, drying one's eyes, holding one's breath in suspense) through metonymic mappings that remove the sense of human agency. That is, the "heaves" are associated with "storm," the form of the verb to wring in "have wrung them" is attached to "Eyes," and it is the "Breaths" that are "gathering firm." Although both eye and breath are metonyms for the watchers in the room, moving them into subject from object position reduces the sense of human agency. The use of the passive "be witnessed" further reinforces the suppression of the agent from the scene. From the dying persona's perspective, human agency is no longer a factor in experience.

What agency there is in the poem is, ironically, encapsulated in the dying persona's actions in the third stanza. But these actions—making a will and signing away (one's rights)—are both indicative of ceding, of giving up the attachment to human things. Even the fly is given reduced agency, through the *there* construction ("there interposed a fly").[8] Nominalization of the verb *buzz* at the beginning to the fly's "Blue - uncertain - stumbling Buzz" at the end shifts the poem from movement to stasis. Most of the verbs are stative—*hear, see, be*—or have patients as subjects—*die, fail*.

The prosodic form of Dickinson's poem is marked by a lack of tension: its iambic pattern is reinforced by stressed syllables falling on the even positions of each line, with only two phrases, "Fly buzz" and "last Onset," diverging (but patterned, with each followed by when clauses: "I died" and "the King / Be witnessed") from this regular

---

[8] Linguistics Professor Thomas Roeper (UMass/Amherst) makes the following observation: "The 'there interposed a fly' is a clever way of sidestepping intentionality. In a sense, the fly goes intentionally where it wants, but it does not intentionally interact with human affairs presumably" (personal communication).

alternation. The ellipses and compressed syntax characteristic of Dickinsonian style are likewise missing in this poem: the sentences are syntactically regular and complete, creating a "flat" style devoid of any emotive spark that might result from clashing images or oxymoronic couplings.

These are just some of the structural patterns that produce the effect of stillness and quietude in this poem. Its persona is assigned/resigned to death, beyond all human feeling, beyond all life-giving forms, beyond all desire or need to believe in life after death, whether or not it exists. The concluding line is one of finality, of cessation of all human activity and feeling. The verb *see* can take two constructions: to see an object or to see in order to perform some task (I could not see to read, to thread a needle). In stating "I could not see to see -," Dickinson's persona has thus ceded from life's activities at the moment of death, as Dickinson creates the forms of language that ironically absent feeling. Only the living can feel. If Dickinson were indeed trying to capture what it is like to experience the moment of death, she has certainly succeeded.

When the processes of mapping successfully capture the intensional elements of a poem, a reader may reach the level of experiencing the hows and whys of a poem's affects and not merely what it is saying. I suggest that this may be a way of evaluating the nature of the mapping that is happening; not simply at the attribute and relational mapping levels, but at a deeper, systematic level. Holyoake and Thagard's (1995) cognitive analyses and my informal experiment (see Chapter 11) indicate that, although we are born with all the cognitive potentialities being human entail, they are not fully activated at birth. The plasticity of our brains develops synaptic pathways as we grow and experience our world and its affordances. The more we learn to apply our cognitive aesthetic faculties of attention, memory, imagination, and judgment, the richer are our experiences. As we learn to listen to the voice of Emily Dickinson's poetry, the more we open up possible pathways of understanding and become sensitive to what lies below and beyond the surface of a poem in coming closer to Dickinson's own cognitive minding.

Conceiving a Universe

As Emily Dickinson was developing her self-awareness as a poet, the middle period of the nineteenth century was a hotbed of tumultuous times. On all levels, intellectual discussions wrestled with questions of morality, marked by such events as the instability of the United States over slavery that led to the Civil War; the work of Sir John Herschel, Charles Darwin, and others on scientific methods over the discovery of evolution; and the emergence of Natural Theology. That Dickinson was well aware of such movements is undeniable, given her breadth of reading and growing interest in the new sciences. What is not so clear is how such events affected her worldview. By cognitively tracing the patterns of her thinking throughout her poetic corpus, we can see how she developed through metaphoric structuring; what I call her "conceptual universe."

Dickinson grew up within a Calvinist, Puritan-dominated milieu. Although in her teens she expressed her desire to be like her friends and undergo Christian conversion, she could not bring herself to do so, nor could she do so when the rest of her family did. Evolutionary theories, developing in the early part of the nineteenth century and culminating in Darwin's much-delayed publication of *Origin of the Species* in 1859, challenged biblical accounts of creation, and cast doubt on conventional views of God's existence. New discoveries in astronomy expanded the notions of a limited universe, and Emerson's growing interest in Oriental philosophies influenced more expansive views of rebirth and transcendentalism. Faced with such heady investigations, Dickinson in her poetry constructed her conceptual universe in rejection of the cognitive principles underlying Calvinist thought by choosing one of two archetypal conceptual metaphors of life in the Western tradition: LIFE IS A JOURNEY and LIFE IS A VOYAGE. These conceptual metaphors appear in linguistic expressions in various ways, such as facing "obstacles" or venturing "abroad" in life. Such conceptual metaphors are based on underlying sensory-motor-emotive processes through various schemas that represent the different ways we bodily interact with our environment. Some are structures of motion, such as BALANCE, CONTAINMENT, PATH, and CYCLE; some orient our bodies deictically, such as NEAR-FAR and CENTER-PERIPHERY; others are structures of force, such as COMPULSION, BLOCKAGE, COUNTERFORCE, RESTRAINT REMOVAL (M. Johnson 1987).

As Lakoff and Johnson (1980, 1998) have shown, the conceptual metaphor LIFE IS A JOURNEY is based on the linear PATH schema, with points along the way:

A< ——————————— >B

Travel along a path can go in either direction, but the LIFE IS A JOURNEY metaphor imposes a one-way direction in time from birth to death. In Christian thought, however, death is not the end of the path, but a point along the way, a gate toward heaven or hell (Figure 13.1).

This is a concept Dickinson could not accept. "Where go we - / Go we anywhere / Creation after this?" she wrote (A223, F1440/J1417/M604). That she rejected the linear LIFE IS A JOURNEY metaphor can be seen in many of her poems. Note, for instance, how, in the following poem, the speaker subverts the journey by first making it incomplete, with the phrase "almost come," then slowing it down because of death's barrier, "the Forest of the Dead," finally to stop altogether with a symbol of surrender in "the white flag" between retreat and God's gates:

> Our journey had advanced -
> Our feet were almost come
> To that odd Fork in Being's
> Road -
> Eternity - by Term -
>
> Our pace took sudden awe -
> Our feet - reluctant - led
> Before - were Cities - but Between -
> The Forest of the Dead -
>
> Retreat - was out of Hope -
> Behind - a Sealed Route -
> Eternity's⁺ White Flag - ⁺ Before - + cool + in front
> And God - at every Gate -

H186, F453/J615/M227

The speaker's discomfort with continuing the journey in this poem is a consistent motif in Dickinson's poems and letters. Her roads are funereal (H119, F722/J735/M361); "the Scarlet way" associated with pain, renunciation, and crucifixion (H64, F404/J527/M215); the speaker on such a road "felt ill - and odd - " (H143, F439/J579/M176); the paths don't so much achieve, or lead to, or even end at, so much as come to a "stop" at their destination (H381, F376/J344/M200):

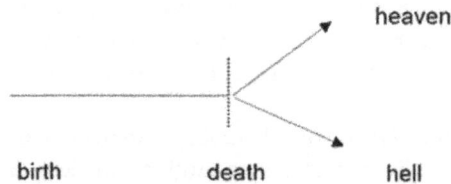

Figure 13.1 The Christian journey of life.

'Twas the old - road -
through pain -
That unfrequented - One -
With many a turn - and
thorn -
That stops - at Heaven -

In an early poem, she mocks the biblical metaphor:

You're right - "the way is
narrow" -
And "difficult the Gate" -
And "few there be" - Correct
again -
That "enter in - thereat" -

'Tis Costly - So are purples!
'Tis just the price of Breath -
With but the "Discount" of
the Grave -
Termed by the Brokers -
"Death"!

And after that - there's
Heaven -
The Good Man's - "Dividend" -
And Bad Men - "go to Jail" -
I guess -

H203, F249/J234/M122

Dickinson explicitly rejects the idea that life is a path that has a specific, predetermined destination. In contemplating the importance of "experience" in our understanding of the world in the following poem, rather than docilely accepting the convention that experience can "lead" us to our destination, she turns it inward into the operations of the mind (H175, F899/J910/M425):

Experience is the Angled
Road
Preferred against the Mind
By - Paradox - the Mind itself -

One significant aspect of the PATH schema is its linear characteristic. The metaphor LIFE IS A JOURNEY is ostensibly grounded in notions of space and spatial orientation, embedded in the notion of "passage." However, since "passage" reflects the notion of time in the aging processes of life, the metaphor is temporally determined by the target domain, LIFE. The word *journey* itself, in its original meaning, meant the distance one could travel in a day (from the French *jour*). More accurately, then, the full metaphoric construct is that of LIFE IS A JOURNEY THROUGH TIME. This point is crucial in understanding Dickinson's rejection of the metaphor, because it was not simply the Calvinist view of life's journey toward heaven that she could not accept; she could not accept traditional notions of time, either. Sometimes she denies clichéd attitudes, as in "They say that 'Time / assuages' - / Time never did assuage -" (H163, F861/J686/M395), or "Death's Waylaying / not the sharpest / Of the Thefts / of Time - " (RML, F1315/J1296/M713); sometimes she condescends to time: "He doubtless did his best - " (A278, F1251/J1478/M559); but it is in her treatment of time in its relation to eternity on the one hand and the world on the other that we see the complexity of her attitudes toward it.

1 Dickinson's Complexities of Time

Cognitive theory claims that thoughts are embodied. That is, we conceptualize our ideas about the world and ourselves through our embodied experience of the world and self. That experience is constrained by the physical orientation of our bodies in time and space, by the constitution of our sense organs, by the repetitive neural synapses of our brains. Abstract ideas, like love, life, and the pursuit of happiness, are understood through the conceptual projection of physical experience. In other words, we cannot think abstractly without thinking metaphorically. Metaphor is not a matter of words but a matter of thought. For instance, we cannot directly experience the notion of time passing. The way we talk about time is conceptually projected from our experience of space, and the expressions we use to describe time are invariably metaphoric. Thus, we say that the past is "behind" us or our future lies "ahead."[1] Time is perceived as a bounded region or container so that we say we are "in" time or "out of" time. We see moments of time from our own vantage point or deictic center, and we can shift that deictic center when we create a mental space that is different from our current "reality" space. Thus, in a sentence like "James plans to leave London to go to Rome tomorrow," a present reality space has been created in which James is making his plans, and a future space in which he will leave his present location to move into another.

How do we understand time? It is commonly understood in two ways, depending on figure-ground orientation. That is, we can perceive time as a figure

---

[1] Such a perspective is culturally determined: "Time seems to come at us head-on: the future in front, the past behind. Not so for the Aymara people of the Andes. Because the past is what they have experienced, it lies ahead, where they can see it. The future remains hidden, so it is behind them. That's because visual evidence is particularly important to the Aymara" (https://www.popsci.com/what-time-looks-like-to-different-cultures/).

with respect to some ground, as when we say "Time flies when we're having fun," where time is seen as passing quickly across some given fun-filled space. Or we can perceive time as the ground for the figure, as when we say "The train arrived on time." Both these ways of looking at time come from a very general metaphor in our thought processes: the EVENT STRUCTURE metaphor. When applied to the concept of time, this metaphor produces the dual metaphors of TIME IS AN OBJECT and TIME IS LOCATION. These dual metaphors produce thousands of metaphors in everyday language. For example, TIME IS AN OBJECT: Do you have time to go over this paper for me? Where did all the time go? TIME IS LOCATION: Where did you pass the time? Did you arrive in time? We are almost out of time. These dual examples also involve the very common schema in our everyday conceptions: the CONTAINMENT schema.

As may be expected, Emily Dickinson, like any other poet, makes use of the full range of systematic metaphors for time, whether as OBJECT or LOCATION. Time can be barefoot (H170, F496/J717B/M246); weighed (A697/A86-7/8, F949/J834/M442); narrow (H191, F1100/J1100/M491); it can come and go; we can be between eternity and time (H44, F713/J644/M348); we can look back on time (A278, F1251/J1478/M559), and so on. But, you may argue, so what? All poets make extensive use of metaphor; that's what poetry is; literary criticism is full of metaphor identification in poetry. Just so. But the question is: Are these metaphors just strategies for enlivening otherwise prosaic language, for making language fresh through the use of novel metaphors? Or are they indications of a systematic pattern, a marker of the poet's way of thinking about the world, a sign of her conceptual universe?

Emily Dickinson is a great poet, not only because she is a skillful wielder of words, but because she understands the metaphorical nature of our everyday language and thought. She makes use of that knowledge to create poems that literally take our breath away by disrupting our commonsensical and folk theory ways of thinking about the world. For example, a very common metaphor for time is TIME IS A HEALER. This metaphor for time depends on the EVENT STRUCTURE metaphor that entails EVENTS ARE ACTIONS, which in turn entails TIME IS AN OBJECT. The EVENT STRUCTURE metaphor is shaped by the notion of causality, in which an agent is understood to bring about an event. Thus we say "Time heals all wounds." But Dickinson rejects this metaphor:

> They say that "Time
> assuages" -
> Time never did assuage -
> An actual suffering
> strengthens
> As sinews do - with Age -
>
> Time is a Test of
> Trouble -
> But not a Remedy -
> If such it prove, it

prove too
There was no Malady -

                                                                    H163, F861/J686/M395

Dickinson denies the folk theory metaphor that TIME IS A HEALER in order to make the point that true suffering is everlasting. She rejects the idea of time as an agentive figure working against the ground of suffering and replaces it by reversing figure and ground. In the second part of the poem, it is suffering or "trouble" that is perceived as the figure against the ground of time. She replaces the TIME IS CAUSATION metaphor with one in which time is perceived as a standard, a criterion by which suffering may be judged. The words "test" and "prove" suggest the methodology of science, by which "actual" experience may be empirically verified. What is being suggested in this poem is that some metaphors are better than others in enabling us to understand life's experiences.

Another poem with similar figure-ground configuration sees time as a bounded region in space that pain expands or contracts:

Pain - expands the Time -
Ages ⁺ coil within   + lurk
The minutest Circumference
Of a single Brain -
Pain contracts - the Time -
Occupied with shot
⁺ Gammuts of Eternities   + Triplets
⁺ Are as they were not -   + flit- show -

                                                                    H211, F833/J967/M410

In this poem, time is not something objective that exists independently of our conceptualization. It is contained within the circumference of our brains and can expand and contract at will. The use of the musical term "Gammuts" (Dickinson's spelling) or, in the variant, "Triplets" relates time to eternities, plural, as if they were the components of eternities.[2] Distinguishing between the meanings of variants is another way of realizing how Dickinson searched in her labyrinthine journey for the "right" word to reflect her minding. The poem ends in a negative hypothetical mental space in which pain denies the existence of these eternities. The relation of the concepts of time and eternity in Dickinson's conceptual universe is difficult to comprehend unless we are willing to forsake our folk theory notion of eternity as an object coming somehow "after" time.

In the poems just quoted, Dickinson rejects the idea of time as an agentive figure working against the ground of suffering or pain. Instead, time is seen as the ground

---

[2] The choice of variant reveals Dickinson's knowledge of music. As Emily Seelbinder (personal correspondence) explains, "Running a gamut is going through an entire scale, playing all the notes, as it were. Triplets are a time-inflection, a disruption of the pattern that hits the ears out of time, but in time with the beats of the measure." Either way, they reflect another dimension to Dickinson's definition of pain as expanding or contracting time.

against which pain and suffering are highlighted. In the following poem, time is again understood as LOCATION, a bounded region encircling or containing the speaker.

> Time feels so vast that
> were it not
> For an Eternity -
> I fear me this Circumference
> Engross my Finity -
> To His exclusion, who
> prepare
> By [+] Processes of size   + Rudiments / Prefaces of size
> For the stupendous Vision   for the stupendous Volume -
> Of His Diameters -
>
> H162, F858/J802/M394

Within a few lines Dickinson has created in this poem a complex series of alternative mental spaces—epistemic, conditional, causative, hypothetical, counterfactual. Without the counterfactual expression, *were it not for an eternity*, the conceptualization is straightforward: the vastness of time results in the speaker's containment within the circumference of her experience. However, these words "sore thumb" this straightforward reading. It is the notion of eternity that enables us to think beyond time.

How we interpret the word "for" depends on the conceptual mapping we, as readers with our own presupposed cognitive mappings, bring to the poem. If we believe that eternity has an objective existence, exterior to time, we will understand "for" to refer to existence, as in "I didn't want for money," with TIME AS OBJECT, so that in the poem eternity is seen as an agent preventing the speaker from thoughts of annihilation. If, however, we believe that eternity is "in" time, as a component of time, with TIME AS LOCATION, we would read the word "for" as durational, as in "I slept for hours," with the result that the fear of annihilation rests on a conditional state—whether time does or does not last (extend) forever. The difference between the two readings is as vast as the time itself that Dickinson describes in the poem: it is the difference between the security of a safe and comforting belief in an afterlife compared with the existential angst that time itself may not be eternal. Dickinson, characteristically, could not rest with the superficial comfort of the first; she could not accept that eternity somehow "kicks in" after time is over:

> Forever - is composed of Nows -
> 'Tis not a different time -
> Except for Infiniteness -
> And Latitude of Home -
> From this - experienced Here -
> Remove the Dates - to These -
> Let Months dissolve in [+] further   + other -

Months -
And Years - exhale in Years -
Without + Debate - or Pause -
+ Certificate -
Or Celebrated Days -
+ No different Our Years would be
+ As infinite -
From Anno Dominies -

H216, F690/J624/M334

Dickinson rejects the notion that eternity lies in some region somewhere in the future that we are not in yet. Rather, it is a space called "forever" that is composed of "nows." It is not a different time from "now," it is present, not future. It is "different" though, unlike those nows, in that it is not bounded but boundless, not finite but infinite, and its region or latitude is "home," where our deictic center is prototypically located. From this vantage point, this deictic center "here," we can take away the "dates" we assign to our "nows." The container of time is a vast sea in which months can dissolve; or a body from which years may be "exhaled." The suggested alternative "other" reinforces the rejection of time as having some future or "further" existence that is different from now.

The final stanza is practically a poetic version of cognitive theory: it is the conceptualizing processes of our cognizing minds that impose discrete periods on time, that divide one moment from another just as we are separated in argument or "debate," that insert "pauses" or identify particular days for "celebration," like Thanksgiving or Christmas, in the ceaseless "flow" of time. The alternative "certificate" for "debate" suggests the stamping of an individuated identity on an otherwise "undifferentiated" time. It is we who impose divisions and identities on time, with the result that we are trapped in our own metaphorizing of the world in which we live.

Dickinson found it difficult, if not impossible, to accept the notion that "death" was at the "end" of a linear progression of a "lifetime" and that "eternity" somehow came after. For Dickinson, eternity was "in time" (H161, F855B/J800/M392). A cognitive approach, by identifying the habitual and coherent conceptual metaphors Dickinson uses, can arrive at a possible understanding of the poet's conceptual world.

2 Metaphors of Space

If Dickinson rejected the LIFE IS A JOURNEY THROUGH TIME metaphor, what did she replace it with? The answer, perhaps, was literally all around her. From the details of nature in its annual cycles, the circumference of hills that surround the valley in which the town of Amherst lies, and, ultimately, from the discoveries of the new sciences, Dickinson transformed the metaphor of LIFE IS A JOURNEY THROUGH TIME into that of LIFE IS A VOYAGE IN SPACE.

It is perhaps difficult for us, with the arrival of space travel, to imagine the concept of previous centuries that the solar system, though large, was not very distant, that earth was judged to be some 4.8 million miles from the nearest star. Not until Bressel succeeded in the first precise measurement, by means of parallax, of the distance from the Earth to a star in 1838, only eight years after Dickinson's birth, did scientists begin to establish just how vast the universe is. New discoveries were happening all the time. Through the work of Thomas Wright and Lambert, disk-shaped galaxies were discovered, with the sun no longer centered but at the edge of the disk that formed the Milky Way. The stars and planets were seen to be afloat in a great expanse, and scientific metaphors developed which saw space as a vast sea, with the planets as boats, circling in sweeps around the sun (Ferris 1988).

Such heady stuff provided Dickinson with the imagery she needed. From her early childhood days at Amherst Academy, Dickinson took a keen interest in developments in all the physical sciences, from botany to astronomy. In his biography of Dickinson, Richard Sewall (1974, 343) describes how Edward Hitchcock, president of Amherst College from 1845 to 1854, was known for his meticulous scientific observations and made Amherst a leading center for scientific study. In his writings, Hitchcock developed a Natural Theology in which he attempted to reconcile a devout belief in revealed religion with the new scientific discoveries of his day. Dickinson's language is full of terminology related to the astronomical achievements of the previous century and to the contemporary events and discoveries happening in her lifetime. This language is not incidental. Dickinson links it imaginatively into a coherent conceptualization realized by the metaphor LIFE IS A VOYAGE IN SPACE. This she achieved through the creation of the metaphor AIR IS SEA that structures the way she conceives her universe.

3 The AIR IS SEA Conceptual Metaphor

Throughout the poetry, sea substitutes for air (A113, F11991/J198/M556):

> A soft Sea washed around the House
> A Sea of Summer Air

In one poem, the comparison is made clear (H106, F469/J484/M234):

> My Garden - like the Beach -
> Denotes there be - a Sea
> That's Summer -

AIR IS SEA is a cognitive metaphor that productively generates linguistic metaphorical expressions to create a coherent view of Dickinson's conceptual universe, as the following examples show.

If AIR IS SEA, then EVERYTHING THAT FLIES IS A SAILOR:

Bags of Doubloons - adventurous Bees / Brought me - from firmamental seas - (H37, F266/J247/M131)

For Captain was the Butterfly / For Helmsman was the Bee (A113, F1199/ J1198/M556)

As Bird's far Navigation / . . . / A plash of Oars, a Gaiety - (H32, F257/J243/M126)

And he unrolled his feathers / And rowed him softer home - (A85-9/10, F359C/J328/M189)

In like manner, action verbs associated with everything that flies are also associated with the sea:

A Sparrow . . . / Invigorated, waded / In all the deepest Sky (A114, F1257*B*/ J1211/M561)

Or Butterflies, off Banks of Noon / Leap, plashless as they swim. (A85-9/10, F359C/J328/M189)

Sometimes, EVERYTHING THAT FLIES IS A BOAT, as in the poem in which a bird leaves its nest for the first time (H161, F853/J798/M392):

> And now, among Circumference -
> Her steady Boat be seen -
> At home - among the Billows - As
> The Bough where she was born -

Even an insect like the "Summer Gnat" is "Unconscious that his single Fleet / Do not comprise the skies -" (H159, F848/J796/M390).

Sunsets, too, are drawn into the metaphor, as air, now understood as sea, is projected onto the sky and ultimately onto space:

> Where Ships of Purple - gently toss -
> On Seas of Daffodil -
> Fantastic Sailors - mingle -
> And then - the Wharf is still!
>
> H110, F296/J265/M139

In one poem, the sunset becomes the sea itself, and images associated with the sea—traffic, landing, bales, merchantmen—constitute the traces left in the sky by the sun in its setting:

> This - is the land - the Sunset washes -
> These - are the Banks of the Yellow Sea -
> Where it rose - or whither it rushes -
> These - are the Western Mystery!
> Night after Night

> Her purple traffic
> Strews the landing with Opal Bales -
> Merchantmen - poise upon Horizons -
> Dip - and vanish like Orioles!
>
> H110, F297/J266/M139

SEA IS AIR also encompasses the human being to give the metaphor THE HUMAN BEING IS A SAILOR:

> Down Time's quaint stream
> Without an oar
> We are enforced to sail
> Our Port a Secret
> Our Perchance a Gale
> What Skipper would
> Incur the Risk
> What Buccaneer would ride
> Without a surety from the Wind
> Or schedule of the Tide -
>
> H ST16e-17a, F1721/J1656/M672

Like everything that flies, THE HUMAN BEING IS A BOAT:

> One port - suffices - for a Brig - like mine -
> (H66, F410/J368/M218)
> If my Bark sink / 'Tis to another sea -
> (A199/200, F1250/J1234/M559)

As with everything that flies, THE HUMAN BEING is also associated with sea-related action verbs:

> I can wade Grief - (H126, F312/J252/M149)
> Sweet Pirate of the heart, / Not Pirate of the
> Sea, / What wrecketh thee? (A42, F1568/J1546/
> M726)

A hint of piracy (from childhood scenes of "walking the plank") hovers around the following poem that describes the voyage of life:

> I stepped from Plank to Plank
> A slow and cautious way
> The Stars about my Head I felt
> About my Feet the Sea.
> I knew not but the next

Would be my final inch -
This gave me the precarious Gait
Some call Experience.

A91-1/2, F926/J875/M434

As this poem suggests, the AIR IS SEA metaphor transforms the "voyage" of life, common to conventional views, from one that is earth-bound to one that takes place within the context of outer space, as the speaker is poised between star and sea. Just as LIFE IS A JOURNEY THOUGH TIME depends on the linear PATH schema, LIFE IS A VOYAGE IN SPACE depends on the circular schema of CYCLE.

4 The CYCLE Schema and the Language of Space and Time

Mark Johnson (1987, 119) describes the CYCLE schema as part of our physiological make-up: "We experience our world and everything in it as embedded within cyclic processes: day and night, the seasons, the course of life (birth through death), the stages of developments in plants and animals, the revolutions of the heavenly bodies." The schema is also something imposed by conventional cycles, such as the time constructs we have created in Western tradition: the hour, the week, the year. Johnson cites four features shared by conventional cycles:

1. cycles constitute temporal boundaries for our activities;
2. cycles are multiple, overlapping, and sequential;
3. cycles can be quantitatively measured according to the mathematics of time, but they will also have qualitative differentiation;
4. there is a difference between "natural" and "conventional" cycles. (120–1).

As Johnson says: "Most fundamentally, a cycle is a temporal circle" in which "backtracking is not permitted" (119).

Dickinson was enormously sensitive to the natural cycles of the seasons, the recurrent change from day to night, the daily routines of the household. However, one way in which her imagination reached beyond the boundaries of the cultural model of CYCLE she inherited was to spatialize its temporal construct. Thus she often changes the linear trajectory of things that move—everything that flies, sun, stars, planets, and human beings—into a circular one:

Butterflies from St Domingo / Cruising round the purple line - (H28, F95/J137/M62)
Within my Garden, rides a Bird / Upon a single Wheel - (H26, F370/J500/M197)
And all the Earth strove common round - (H209, F826/J965/M407)
Convulsion - playing round - / Harmless - as streaks of Meteor - (A679, F187/J792/M376)

The Feet, mechanical, go round - / Of Ground, or Air, or Ought - (H26, F372/
J341/M198)

I worried Nature with my Wheels / When Her's had ceased to run - (A81-10/11,
F887/J786/M403)

To these components we can add the seasons, the weather, life itself:

[Autumn] eddies like a Rose - away - / Upon Vermillion Wheels - (H105, F465/
J656/M233)

The Seasons played around his knees (H176, F970/J975/M449)

As Floods - on Whites of Wheels - (H97, F739/J788/M370)

Of Life's penurious Round - (H43, F283/J313/M346)

To take just one of the many metaphors in the poetry that deal with space and that are generated from the CYCLE schema, I choose the obvious one of the cycle itself, which Dickinson invariably associates with the word *wheel*.[3] The word *cycle* in Dickinson's poems refers not to a static circle but to the cyclical movement of the planets and is, as expected, associated with time. The addition of *wheel* creates the metaphorical extension of movement through space. In an early poem in which Dickinson asks God to find a place for the mouse that has been killed by a cat, she imagines it "Snug in seraphic Cupboards" while "unsuspecting Cycles / Wheel solemnly away!" (H7, F151*B*/ J61/M90). In a somewhat more serious poem, again on the subject of death, in which eternity is associated with sea imagery, the speaker imagines time as a movement through space (H202, F132*B*/J160/M120):

> Next time, to tarry,
> While the Ages steal -
> Slow tramp the Centuries,
> And the Cycles wheel!

Dickinson found the schema of CYCLE more productive than the schema of PATH because it accorded more closely with her conception of the physical world. Although Johnson identifies the CYCLE schema with time, it is also closely associated, as Dickinson saw, with the movement of the earth in space.

The notion of an infinite universe, encouraged by the discoveries of the new science, enabled Dickinson to relate both the temporal and spatial elements of her geography and the AIR IS SEA metaphor in a mapping of the details of the particular world around her into the vaster world of space. Just as the etymology of the word *journey* includes the element of time, as we have seen, so the etymology of the word *voyage* includes the element of space, since the morpheme *voy* comes from the Latin *via*, meaning path or

---

[3] Lest readers presume this to be no metaphor, remember that bicycles were not invented until later in the century; the word *cycle* comes from the Greek *kuklos*, which also means wheel, a point Dickinson would have recognized and appreciated from her beloved lexicon.

```
                    north
                    winter
                   midnight
                    white
                     air


    west                                      east
    fall                                      spring
    evening,                                  dawn, sunrise
    sunset                                    [no one color]
    purple                                    desert/wilderness


                    south
                    summer
                     noon
                     red
                     fire
```

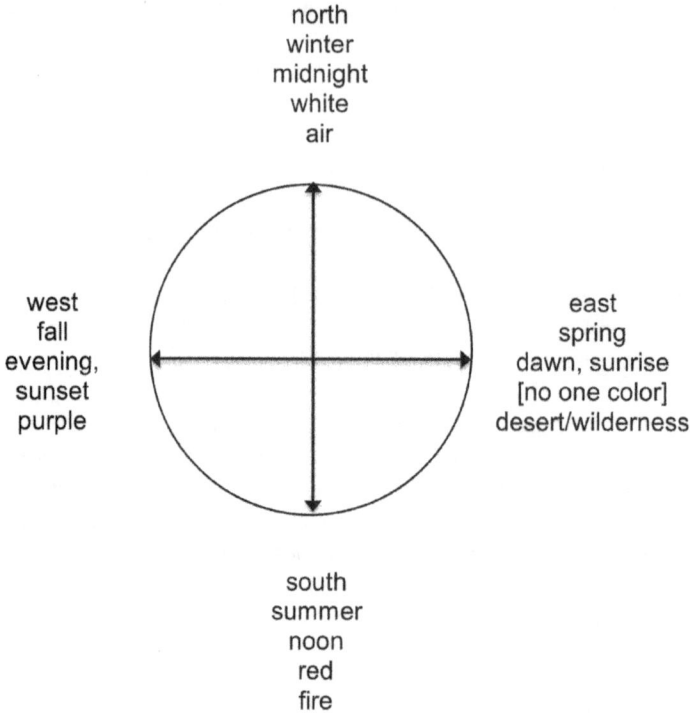

Figure 13.2 Dickinson's cardinal points (based on Patterson 1979).

way. By revealing the metaphorical TIME relation to the processes of life in the SPACE schema that underlies the schema of PATH in the JOURNEY metaphor and by revealing the metaphorical SPACE relation to the physical universe in the schema of TIME that underlies the schema of CYCLE in the VOYAGE metaphor, Dickinson created a world view in which physical location and temporal constructs come together, as earth's four compass points are aligned with the four seasons, times of day, earth's elements, and even colors (Figure 13.2).

The points on the compass where the lines bisect the circle represent the climaxes of the CYCLE schema that we impose on the life cycle we experience, in Johnson's (1987, 120) observation, "as moving from birth to the fulness of maturation followed by a decline toward death." By overlapping the temporal constructs of the daily and annual climaxes with the geographically determined points of the compass, Dickinson has transformed them into spatial schemas. Thus, temporal climaxes on the circle are mapped on to the source domain of space, so that TIME IS LOCATION:

A Music numerous as space - / But neighboring as Noon - (A81-8/9, F504*B*/J783/
     M402)
Whose galleries - are Sunrise - (A796, F208*A*/J161/M135)

To continents of summer - / To firmaments of sun - (H17, F177/J180/M103)
Winter under cultivation / Is as arable as Spring (H ST16d, F1720/J1707/M672)
Besides the Autumn poets sing (A654, F123/J131/M82)

It could, perhaps, be argued that Dickinson was simply using the image schemas of both PATH and CYCLE, which Lakoff and Johnson have shown are basic to conventional interpretations of experience. With respect to her understanding of life, death, and immortality, however, one can see from her very earliest poems that this is not the case: that what she did, in fact, was to replace the PATH schema of conventional attitudes toward immortality (her "Flood subject") with the CYCLE schema projected onto her understanding of space, time, and the universe. The tensions between the two schemas can be seen to be developing, for example, in the following poem in which the sea invocation of "strait pass" is placed, right at the center of the poem, within a planetary scene, as "Convulsion" (itself a turning, a convolution) plays "round" the straight (the spelling in another copy) path, and the martyrs' "Expectation" is turned into an image of a compass needle wading through "polar air":

> Through the strait pass
> of suffering -
> The Martyrs - even - trod.
> Their feet - upon Temptation -
> Their faces - upon God -
> A stately - shriven -
> Company -
> Convulsion - playing round -
> Harmless - as streaks
> of Meteor -
> Opon a Planet's Bond -
> Their faith -
> the everlasting troth -
> Their Expectation - fair -
> The Needle - to the North
> Degree -
> Wades - so - thro' polar Air!

A679, F187B/J792/M376

Even though the martyrs believe they are proceeding on a linear path that will lead them to their final destination, in the planetary scene of Dickinson's world, paths are in fact orbits, with the result that what seems straight to the martyrs is in fact circular and cyclical. Their faith and their expectation, the "everlasting" covenant they have made with God, is ironically compared with the needle of the compass pointing to magnetic, not true, north, just as the poles, which mark the diameter of the earth's sphere, appear fixed but are actually moving in space, kept in their orbit by the sun's "Bond."

In what is perhaps her most famous "journey" poem, "Because I could not stop for Death - " (H165, F479/J712/M239), the same pattern of change at the very center of the poem occurs as the journey is abruptly terminated with a cyclical image of movement in space: "We passed the Setting Sun - / Or rather - He passed Us -," and the tone changes from a pleasant afternoon's ride to the "quivering and chill" of the grave. This stanza (interestingly omitted from the poem on its first publication) transforms the poem from an otherwise fairly orthodox account of life's journey to one that is more problematic and foreshadows the incompletion at the end, as time and the journey stand still:

> Since then - 'tis Centuries - and yet
> Feels shorter than the Day
> I first surmised the Horses Heads
> Were toward Eternity -

In "No Man can compass a Despair -" (H44, F714/J477/M349), the poet compares the man to a traveler going round "a Goalless Road" who is "Unconscious of the Width - / Unconscious that the Sun / Be setting on His progress -." With such a metaphorical restructuring of the linear, temporal characteristic of the journey into a circular, spatial orientation, Dickinson formulated a vision of a world in which the dead have no place. Unlike religious interpretations of the LIFE IS A JOURNEY THROUGH TIME metaphor, in which the afterlife is its destination and death merely a gate on the way (Figure 13.1), Dickinson's new metaphor had no place for either in her conceptual universe (Figure 13.3).

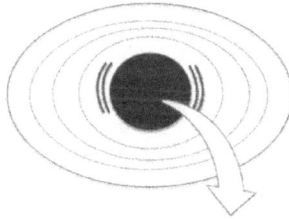

Death ?

Figure 13.3 Spatial location of death in a cyclical universe.

In a cyclical universe, the geographical metaphors of goal, location as "up" or "end" have no physical, bodily grounding, with the consequence that it no longer makes sense to speak of "destination" after death. The problematic "location" of the dead is raised in the following poem which confronts the conflict between the two metaphors directly as astronomy replaces revealed religion in establishing God's existence:

> The Moon upon
> her fluent Route

Defiant of a Road -
The Star's
Etruscan Argument
Substantiate a God -
If Aims impel
these Astral Ones
The ones allowed
to know
Know that which
makes them as
forgot
As Dawn
forgets them -
now -

A806, F1574C/J1528/M636

If there is purpose in the universe, only the dead can know—but they are beyond living memory and time. A variant to the last four lines of this poem makes even clearer the loss of heaven as an anticipated goal at the end of life's journey with the shift to a cosmological perspective and the disruption of linear time:

How archly spared the Heaven "to come" -
If such prospective be -
By superseding Destiny
And dwelling there Today -

With the LIFE IS A VOYAGE IN SPACE metaphor, Dickinson conceives a new place for life on earth. It is almost as though Dickinson has anticipated the concept of modern physics that not just the world but the universe itself can be finite but unbounded. No longer are we travelers on life's road, but we are identified with the earth itself in its daily rotation, as the first and last stanzas of the following poem show (H99, F743/J721/M373):

Behind Me - dips Eternity -
Before Me - Immortality -
Myself - the Term between -
Death but the Drift of Eastern Gray,
Dissolving into Dawn away,
Before the West begin -
Tis Miracle before Me - then -
[. . .]
'Tis Miracle behind - between -
A Crescent in the Sea -
With Midnight to the North of Her -

And Midnight to the South of Her -
And Maelstrom - in the Sky -

Dickinson contemplates the seemingly infinite reaches beyond the solar system as she defines eternity in terms of space:

As if the Sea should
part
And show a further Sea -
And that - a further - and
the Three
But a presumption be -
Of Periods of Seas -
Unvisited of Shores -
Themselves the Verge of
Seas to be -
Eternity - is Those -

H119, F720/J695/M361

Infinity itself is seen as a giant extending across the diameters of the earth, as one poem begins with infinity and ends with the vital life ("quick") of eternity:

They leave us with the
Infinite -
But He - is not a man -
His Fingers are the size
of Fists -
His Fists, the Size of
Men -
And whom he foundeth
with his Arm
As Himmaleh - shall
stand -
Gibraltar's Everlasting
shoe
Poised lightly on his Hand -
So trust Him - Comrade!
You for you - and I -
For you and me -
Eternity is ample -
And quick enough -
If true -

A470/A85-3/4, F352A/J350/M185

"How infinite - to be / Alive - " Dickinson exclaimed (A84-9/10, F605/J470/M276). It might be:

> Finite - to fail, but
> infinite to Venture -
> For the one ship that
> struts the Shore
> Many's the gallant -
> overwhelmed Creature
> Nodding in Navies
> Nevermore -
>
> A86-9/10, F952/J847/M443

Perhaps one of the most revealing poems that suggests "The Finite - furnished / With the Infinite" (H211, F830/J906/M409) is "By my Window have I for Scenery / Just a Sea - with a Stem -" (H160, F849/J797/M390), which unites the metaphors of air/sea with those of space, ending with an explicit reference to the pine tree as a member of scientific establishment in the final stanza:

> Was the Pine at my Window a "Fellow
> Of the Royal" Infinity?
> Apprehensions - are God's introductions -
> To be hallowed - accordingly -

The voyage of the soul is an embarkation on eternity's sea (H254, F143/J76/M81):

> Exultation is the going
> Of an inland soul to sea,
> Past the houses - past the headlands -
> Into deep Eternity -

That Dickinson personally identified with her metaphor can be seen in her correspondence. To her presumably ill sister-in-law she wrote: "You must let me go first, Sue, because I live in the Sea always and know the Road" (L306). Life lived in Dickinson's metaphorical sea is graphically portrayed in the following poem:

> Escaping backward to
> perceive
> The Sea opon our place -
> Escaping forward - to
> confront
> His glittering Embrace -
> Retreating up, a Billow's
> hight

> Retreating blinded down
> Our undermining feet
> to meet
> Instructs to the Divine -
>
> A87-7/8, F969/J867/M448

Infinity in time, eternity in space. Had Dickinson known about the space-time continuum and black holes, in which time and space exchange places, she might have found the metaphor she was searching for that would enable her to unite life and death. As it was, she was left suspended at the point in which the old order had been discredited by the new science, but more questions had been raised as a result. She would, however, have understood Freeman Dyson's lectures in *Infinite in All Directions* (1980), and perhaps seen her own attempt to link the things of this world to the universe in the images he uses:

> Butterflies are at the extreme of concreteness, superstrings at the extreme of abstraction. They mark the extreme limits of the territory over which science claims jurisdiction. Both are, in their different ways, beautiful. Both are, from a scientific point of view, poorly understood. Scientifically speaking, a butterfly is at least as mysterious as a superstring. (14)

That is a concept Dickinson would have understood. Poets have, through the ages, been credited with the ability to speak truths, to capture, somehow, the "truths" of the universe through a different path from the ones scientists take. "Tell all the Truth but tell it slant - / Success in Circuit lies" (A372, F1263/J1129/M563) was Dickinson's way of putting it. In attempting to describe what poets do, however, we reach the "fudge factor" when we try to explain how poets "tell truths"; how their work somehow illuminates for us the nature of the world and the nature of human understanding. We fail to do so when we impose the false theoretical construct of "objective reality" on physical—and poetic—reality.

To read cognitively, therefore, is to take into consideration Dickinson's cognitive responses to the historical contexts in which she was writing and to perceive how her conceptual metaphors reflect current scientific knowledge or refute traditional presuppositions underlying the way we think about our lives and experiences of space and time. By recognizing the productive cognitive metaphors that structure a poetics, we can see how they enable a poet like Dickinson to create her own individual world truth, a truth that is grounded in a physically embodied universe. Her choice of the CYCLE over that of the PATH schema substitutes circularity for linearity and thus opens her poetics to the underlying schemas of CONTAINMENT and TRANSFORMATION.

A Transformative Poetics

I find ecstasy in living—the mere sense of living is joy enough
—Dickinson's comment to Higginson (L342a).

It is often remarked that Dickinson said little if anything about the principles on which her ideas of poetry are founded. This is not quite true. Although she never wrote a treatise, she didn't need to. Her poetics were summed up succinctly in three brief sentences Higginson recorded her as saying:

> If I read a book [and] it makes my whole body so cold no fire ever can warm me I know *that* is poetry. If I feel physically as if the top of my head were taken off, I know *that* is poetry. These are the only way I know it. Is there any other way. (L342a)

In an earlier letter to Higginson, she wrote: "I had no Monarch in my life, and cannot rule myself, and when I try to organize—my little Force explodes—and leaves me bare and charred" (L271). These quotes reveal Dickinson's poetics from the viewpoint of both poet and reader.

Dickinson's remarks show that reading a poem that "works"—that accomplishes what a poet sets out to do—creates sensory, visceral, motor, and emotive responses, and that the act of writing a poem is triggered by the force of such subliminal motivations. Both quotes speak to the sensory, motor, and emotive feelings that writing and reading poetry evoke and indicate the cognitive principles by which poets formulate their poetry. The ideas of both poets and readers as to what poetry is and what it is for them are founded, not on conceptual interpretive reasoning, but on the subliminal, "gut" feelings experienced by and arising from the motivating forces of sensory, motor, and emotive (per)ceptions.

In pointing out the problematic boundary separation between "perception" as sensate and "conception" as cognitive, Leonard Talmy (2000, 139) writes:

> Accordingly, it seems advisable to establish a theoretical framework that does not imply discrete categories and clearly located boundaries, and that recognizes a cognitive domain encompassing traditional notions of both perception and conception. . . . To this end, we adopt the notion of ception here to cover all the

cognitive phenomena, conscious and unconscious, understood by the conjunction of perception and conception. While perhaps best limited to the phenomena of current processing, ception would include the processing of sensory simulation, mental imagery, and currently experienced thought and effect.

These "ceptions" are created and articulated through schemas that Immanuel Kant (1934[1787]) notes "lie at the foundation of our pure sensuous conception."

Mark Johnson (1987, 23–4) describes schemas as connecting physical, kinetic, sensory, and emotive perceptions to their conceptualizations in language. He is careful to clarify that what he calls image schemas "*are not rich, concrete images or mental pictures either. They are structures that organize our mental representations at a level more general and abstract than that at which we form particular mental images.*"

Dickinson's poetics are governed by two primary schemas that underlie the structure of many of her productive metaphors: CONTAINMENT and TRANSFORMATION. Johnson identifies five characteristics that are entailed by the CONTAINMENT schema, based on an underlying *in-out* orientation:

1. it typically involves protection from, or resistance to, external forces;
2. it limits and restricts forces within the container;
3. it gives the contained object relative fixity of location;
4. it makes the contained object either accessible or inaccessible to an observer;
5. it has relational property, such that if A is in B and B is in C, A is in C. (22)

We can put things into or take things out of a container, or enter or leave a contained space, such as a house or a room. Our bodies themselves are conceived as containers. Thus, in Dickinson's first sentence, the effect of reading—taking in—poetry makes the "whole body" a container of a coldness resistant to the external pressure of a warming fire.

Affective schemas are structured by the kinesthetic movement and causation of force dynamics. That is, movement can be enabled and allowed by removal of restraint or resisted and impeded by the imposition of restraint (Johnson 1987, 42–53; Talmy 2000, 409–70). CONTAINMENT can be either open, allowing passage in and out, or closed, entailing the necessity of forced escape. When Dickinson says that reading poetry makes her feel "physically as if the top of [her] head were taken off," her body is conceived as an explosive container against the build-up of emotive pressure inside it (Kövesces 1986, 2000). This emotive force occurs also in writing poetry as Dickinson's response to the attempt to control results in a dynamic explosion of emotion and sensory feeling, leaving the poet "bare and charred."

For Dickinson's poetics, the corresponding schema associated with the force dynamics of the CONTAINMENT schema is TRANSFORMATION. Its structure of change from one state to another is, roughly,

1. an initial state;
2. some causative agent acting upon that initial state; and then
3. an emergent or resultative state.

It can be a metamorphosis, when a material object transmutes into another, as an acorn becomes a tree, or a caterpillar becomes a butterfly. It can refer to any change of appearance, character, or state, even the transition from life to death.[1]

In Dickinson's poetry, the two schemas of CONTAINMENT and TRANSFORMATION interrelate in such a way that movement either in or out of the CONTAINER triggers not just a change of place, as in the PATH schema, but a change of form. For Dickinson, given the *in-out* movement entailed by the CONTAINMENT schema, its boundaries exist only to be transgressed. Dickinson's poetry is full of images of containment and breaking boundaries, of an orientation that is either in or out of life's circumference. It is the difference between agency that asserts control and agency that breaks free of control.

The CONTAINMENT/TRANSFORMATION schemas in Dickinson's poetry are entailed by the circular schema of CYCLE, associated with the metaphor LIFE IS A VOYAGE IN SPACE (see Chapter 13). This schema is very different from the linear PATH schema that underlies the LIFE IS A JOURNEY THROUGH TIME metaphor. Unlike our position "on" a straight path, we are either contained inside (within) or outside (without) the boundary of the circular movement of CYCLE. The double meaning of *physically* being inside/ outside or *perceptually* within/without raises the schematic relationship of presence and absence within and without the boundaries of a CIRCLE. Thus, it is possible to have metaphorical projections based on the IN-OUT orientation of the CONTAINMENT/ TRANSFORMATION schemas.

For Dickinson, the CONTAINMENT schema has two forms: closed and open. Only death can ensure a closed container, one that is static, safe, as in *Doom is the house without the door* (H42, F710/J475/M346). In life, the container is either an open one or one that may be breached under pressure. In either case, the open or breached container provides the environment for the force dynamics of movement and transformation to occur, whether it is in the "building of the soul" in *The props assist the house* (H122/ H347, F729/J1142/M365) or the emergence of the butterfly from its cocoon in *My cocoon tightens* (H189, F1107/J1099/M494). The schema of TRANSFORMATION is therefore associated with the CONTAINMENT schema of life, and is dynamic, involving movement and transformation.

The following sections show how the two schemas work in Dickinson's poetry, followed by a more detailed analysis of two poems that encapsulate the way the schemas of CONTAINMENT and TRANSFORMATION underlie and in-form Dickinson's poetics.

---

[1] In his article on Eavan Boland's poetry, Nigel McLoughlin (2017, 199) identifies the following possibilities: "Boland uses metamorphosis as metaphor in several ways. The first of these may be grouped broadly as 'metamorphosis is diminution, decay, reversal or loss of power'; the second may be grouped broadly under 'metamorphosis is usurpation, empowerment, or move towards the place of power'; the third may be classified as 'metamorphosis is renewal or repetition'. Several of the metaphors are complex and may include features of more than one of the basic types."

1 CONTAINMENT as Confinement

For Dickinson, our entire world is a CONTAINER in which we reside. The world can entail confinement, as in the "magic prison" of life portrayed in *Of God we ask one favor* (A819, F1675/J1601/M655). Confinement can sometimes provide a sense of security, as in *Safe in their alabaster chambers* (H11/H203, F124/J216/M83,122). In her extensive study of the way bodily confinement is represented in literature, Monika Fludernik (2019) explores Dickinson's poems referring to prisons. Fludernik distinguishes between those poems that consider the self-incarceration of HOME AS PRISON from the liberation for creativity occasioned by PRISON AS HOME (234–6). In discussing two poems, *A prison gets to be a friend* (H103, F456/J652/M229) and *They shut me up in prose* (H182, F445/J613/M223), Fludernik focuses on both the ambivalence of Dickinson's poetics of incarceration, with its fantasies of freedom (240–3) and how the restrictions on women lead the poet to "an explosion of poetic achievement" (586–90).

Fludernik provides extensive and illuminating analyses of many of the prison metaphors in Dickinson's poetry. In reviewing literary critical analyses of Dickinson's prison metaphors in letters as well as in poems, she argues for readings beyond simply the HOME AS PRISON/PRISON AS HOME metaphors, including religion and death:

> The poet's phobias about enclosure and imprisonment are more than offset by her repeated scenarios of transcendence, repose, and joyful purposiveness, often (whether marked or not) in the language of religious martyrdom. At the same time, the poems also portray a varied selection of different stages of abandonment, despair, and depression, as well as prostration with fear and agony—feelings taken to be autobiographically relevant. (237)

Fludernik's focus is thus on the way Dickinson's use of the poetics of confinement reflects the biographical, psychological, religious, and feminist implications revealed in her "prison" poems (233–43; 586–91). My focus is different; that is, rather than focusing on the way the poetry speaks to real-life concerns, I focus on the way Dickinson transforms real-life concerns into a poetics that transcends those concerns into an overarching and comprehensive account of how poetry works in the world, and how readers of her poetry experience that transformation. In other words, rather than focusing, as Fludernik does, on Dickinson's "poetics of confinement," I focus on how Dickinson's metaphors of confinement contribute to her poetics. Given the dual nature of Dickinson's schematic poetics, I argue that it reflects the poet's overarching topic of escape from CONTAINMENT to TRANSFORMATION. In doing so, Dickinson's poetry iconically reaches beyond the confines of language into the realm of the ineffable.

2 TRANSFORMATION as Escape into the Beyond

The sixteen poems that Fludernik cites explore the notion of restriction on one's bodily freedom from different aspects, focusing on the affective advantages and disadvantages of existing within enclosed space. Although half the prison poems cited by Fludernik can be read as positive or negative, they are balanced by the eight poems that deal with escape (see Appendix). Taken within the context of the poems as a whole, the CONTAINMENT/TRANSFORMATION schemas go well beyond simply the notion of human physical imprisonment to questions of the self, whether human or otherwise, in interaction with the world and others. Typical examples include poems related to the soul, to reading, to questions of freedom, infinity, and eternity, to death itself.

For instance, the idea of the imprisonment of the soul is represented in *The soul has bandaged moments* (A85-11/12, F360/J512/M190). The poem's six stanzas create a containment 2-2-2 structure, whereby the moments of the soul are described in the first two stanzas with a lack of movement as "too appalled to stir, followed by the soul's moments of escape" in the middle two stanzas, only to be "retaken" to prison in the last two. It is those middle two stanzas centered within the containment of the outer two that reveal Dickinson's poetic schemas:

> The soul has moments of Escape -
> When bursting all the doors -
> She dances like a Bomb - abroad -
> And swings opon the hours -
>
> As do the Bee -delirious borne -
> Long Dungeoned from his Rose -
> Touch Liberty - then know no
> more -
> But Noon - and Paradise -

Whether the confinement be negative, as the soul confronts "Fright" and "Horror" in the surrounding stanzas of CONTAINMENT, or positive, as the bee is "Long Dungeoned" while enjoying the rose's nectar, it is the breaking of the boundaries of that confinement that leads to freedom and bliss.[2]

Perhaps her most famous poem to express the two schemas as they participate in the emotion metaphors of explosive force is the following one that describes the refining of the soul in terms of a blacksmith's forge:

> Dare you see a Soul at the
> "White Heat"?
> Then crouch within the door -
> Red - is the Fire's common tint -
> But when the [quickened / vivid] Ore

---

2 This movement from CONTAINMENT to TRANSFORMATION is captured in Lesley Dill's opera "Divide Light," based on Emily Dickinson's poetry, as it progresses from the beginning with *I am afraid to own a body* (H193, F1050/J1090/M472) to the end with *But leave me Ecstasy* (A776, F1671B/J1640/M654).

Has sated Flame's conditions -
She quivers from the Forge
Without a color, but the Light
Of unannointed Blaze -

Least Village, boasts it's Black -
smith -
Whose Anvil's even ring
Stands symbol for the finer Forge
That soundless tugs - within -

Refining these impatient Ores
With Hammer, and with Blaze
Until the designated Light
Repudiate the Forge -

A162, F401A/J365/M214

The form of the poem is circular, not linear, as the second set of two stanzas restates the metaphor of the first set in parallel imagery. The poem introduces the metaphor in the first two stanzas by inviting the reader to look inside the soul as if it were the product of a blacksmith's forge. The soul's transformation is described in terms of the refinement that takes place as the blacksmith's work on the ore's material causes the redness of the fire's flame through the "unannointed" blaze to become the "designated" light of "White Heat."

For Dickinson, poetry is the means by which we are transformed by the possibilities of moving beyond the bounds of circumference in both space and time. The CONTAINMENT/TRANSFORMATION schemas are pervasive throughout Dickinson's poetry, whether expressed in the man-made images of houses, rooms, and graves or in the cocoons, skies, and flora and fauna of the natural world. Her poetry thus becomes an icon that accesses the "something else" beyond itself (M. Freeman 2020).

3 Iconic Realization

According to Peircean semiotic theory, the icon has a complex structure, composed of image, diagram, and metaphor. Not much had been attempted to determine how these might iconically interrelate with metaphor in the literary arts until Masako Hiraga's (1998, 2005) groundbreaking studies on applying Fauconnier and Turner's (2002) blending theory to the role of iconicity and metaphor in poetic texts. From a cognitive perspective, the creation of an icon emerges from the metaphorical triggering of elements from image (*cept*) and diagram (*struct*). In other words, when the isomorphic structures from the two input spaces in the generic space are metaphorically related, the possibility for creating an icon occurs (Figure 14.1).

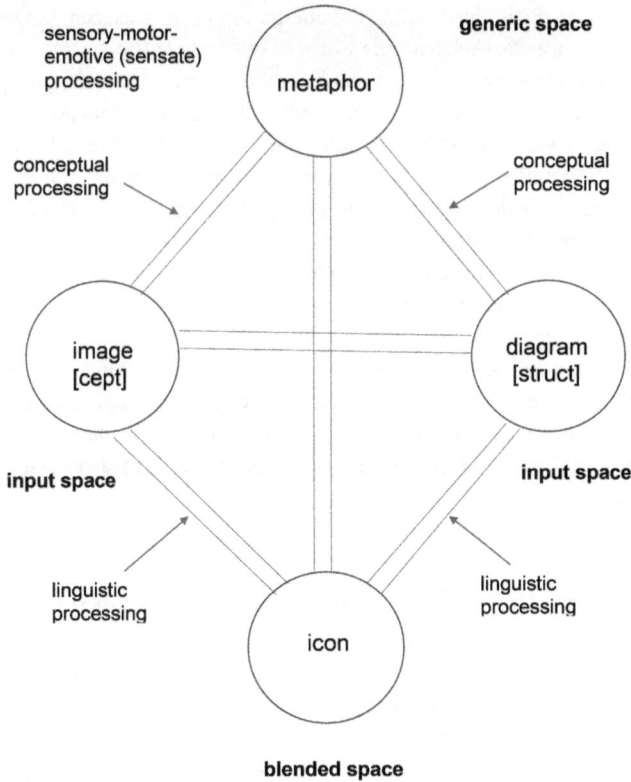

Figure 14.1 Icon blending in cognition.

A metaphor is triggered when the meaning of one expression is transformed into another. Dickinson refers to this explicitly in her poem *We dream* (H92, F584/J531/M266), when she plays with the different meanings of dying:

> What harm? Men die - Externally -
> It is a truth - of Blood -
> But we - are dying in Drama -
> And Drama - is never dead -

When conceptual metaphor governs the process of blending, its emergent structure becomes an icon of experienced reality. For poetry, then, when a poem is governed by a metaphorical structure, the possibility for iconic realization occurs. The poet, from the context of living world experience and motivated by the dynamic schemas that underlie metaphoric conceptualization, creates a poem as the emergent structure of blending elements from the two source and target spaces, thus creating the poem as an icon. The reader, in experiencing the poem's affects, cognitively reconstructs the processes in reverse. By understanding the poem's significance as icon, the reader then projects the inferences that relate to the world of self and context.

A poetic icon occurs when an art object or event—performance, picture, poem, sculpture—successfully invokes something beyond itself. This "beyond itself" is not simply a representation of external reality, nor is it transcendence to the domain of the unreal. It is rather revelation: an attempt to reach beyond the conventional immediacy of our everyday experience, to plunge into and manifest the depths of our primordial experience of precategorial being that underlies the structure of reality both in ourselves and the world of which we are a part. This reality is in-visible to us, not as absence or void, but as being hidden in the visible but always present in the moment (Merleau-Ponty 1968). At these moments we are made, if only momentarily, self-aware of our being as part of primordial being, establishing the "presence of absence." This is the reality that the arts attempt to capture through their emphasis on the concrete, the particular, the individual. By restoring the primacy of the sensory-motor-emotive underpinnings of our experiences, feeling is invested with form and form with feeling.

4 The Role of Metaphor in the Poem as Icon

Metaphorical mapping is integral to the way we perceive and construe our experience of the world. Poetic metaphor is therefore not simply a feature that renders imagery more vivid, nor does it exist only at the conceptual level. It is a structuring process that fuses or blends ceptions at every level of our cognitive functioning, from the subliminal, preconscious sensory, motor, and emotive processes to conscious awareness.

In discussing Dickinson's poem *When Etna basks*, I first describe its relevant metaphors and then show how, through an active process of blending, we access its underlying experience of sensate cognition so that the poem becomes an icon for fear (Figure 14.2).

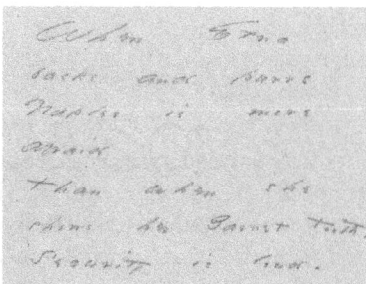

When Etna
basks and purrs
Naples is more
afraid
than when she
shows her Garnet tooth.
Security is loud.

Figure 14.2 *When Etna basks and purrs* (H374, F1161/J1146/M705) ms_am_1118_3_374_0001 Courtesy Houghton Library, Harvard University.

This short poem raises three questions that have puzzled critics and readers:

1. Why is Naples in Italy introduced and not, say, Palermo in Sicily?
2. Why is Naples more afraid of the volcano before rather than when it erupts?
3. Why is Security loud?

Samuel R. Levin (1971) analyzes Dickinson's poem from a propositional-based approach that focuses on grammatical analysis. He argues that:

1. deletions are permissible in grammar only when they are recoverable;
2. compression in Dickinson's Etna poem results from nonrecoverable deletion.

Levin uses feature analysis to discover an antonymic relation between "basks and purrs" and "shows her Garnet Tooth." He claims that "shows her Garnet Tooth" contrasts with "basks" rather than "purrs" because they are opposed both by the same shared features of (+/- Action) and (+/- Contentment) whereas "purrs" is (+ Sound). Because Levin adopts a distributive meaning between "basks" and "purrs" (purring as a result of basking) instead of seeing them as a unified entity, he claims that the lines require a fourth term to provide grammatical symmetry. Levin proposes, given the metaphor of the volcano as a (big) cat, the possibility of the missing term being *hiss, roar,* or *growl.* Levin then moves to the final line of Dickinson's poem: "Security is loud." Again, a feature analysis provides the missing term to provide the needed grammatical symmetry. *Loud* (- soft) contrasts with *purrs* (+ soft).

Noting that the comparison he has been syntactically analyzing is embedded in the sentence *Naples is . . . afraid,* Levin concludes that "although the poem does not make it explicit, this entire structure, including the comparison, is a result clause dependent on the causal clause *Security is loud*" (45). He therefore suggests that a causal subordinate conjunction between the two sentences of the poem has been deleted and provides a syntactic tree for the poem on page 194 (Figure 14.3).

Levin sees the *when* clauses as the major comparison of the poem. To make "explicit the logical relation between the two main clauses of the poem," he introduces a deleted causal conjunction, *because,* to connect them. The effect of this syntax is to make Naples more afraid **because** security is loud. Is this what the poem is asserting?

Although grammatical analysis is an important component of poetic analysis, Levin's reliance on syntactic analysis alone cannot account for the poem's affect. In arguing that linguistic analysis should direct poetic analysis to distinguish the poetic function from normal language use, Levin equates intuitions with linguistic competence. The "stock of intuitions" he names that we have about poetry—unified, compressed, novel—are not the kind of intuitions that we have in experiencing a poem. Rather, they are abstract formulations that *presuppose* the existence of a poetic function. Relying upon a syntactic analysis alone in responding to the poem is limited because

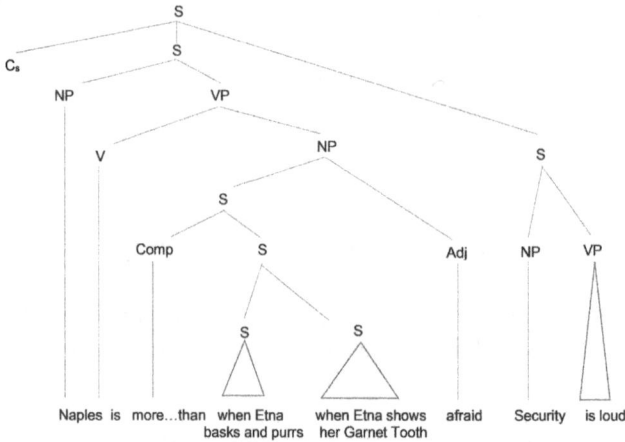

Figure 14.3 Levin's syntactic tree for *When Etna basks and purrs*.

it creates unnecessary and unresolvable ambiguities. Although Levin recognizes the poem's metaphor of volcano as cat, his analysis nowhere considers the metaphor as a structuring principle for the poem.

By metaphorically mapping the actions of an active volcano onto those of a cat, Dickinson compresses our experience of a volcanic eruption, either physically embodied or mentally conceived, with the more intimate human scale of relating to another animal. The activities of a cat basking, purring, and showing her teeth are not put in antonymic or distributive relation, as Levin would have it, but are topological elements in a CAT mental space that the poem makes functionally equivalent to topological elements in the VOLCANO space. There are two different cat metaphors in the poem. The first makes us think of a domestic cat, basking in the warm sunshine and purring. Its effect is of something friendly and non-threatening. The second cat metaphor is of a wild feline showing its teeth. This cat is threatening and dangerous (Figure 14.4).

The two cat metaphors are not comparative but transitional. The two *when* clauses set up a temporal frame that indicates movement through time. This sense of movement is echoed by the transformation of the domestic cat into the ferocity of a wild animal baring its bloody teeth.

The term *garnet tooth* is itself the result of metaphorical blending. Garnet is a silicate dark red mineral that is a key ingredient in marking the temperature-time genesis of igneous rock, as a volcano erupts and its molten lava flows solidify. Animal teeth develop from alveolar bone as they erupt to form various tissue layers of density and hardness. The eruption of both volcanoes and teeth is a fortuitous identity connection that enables compression in the blend (Figure 14.5).

Dickinson's cat metaphors may have been drawn from her literary experience. In the Aeneid, Virgil (1910) describes an Etna eruption. Virgil vividly depicts the fiery heat and loud sounds of eruption in terms that portray Etna as a wild and monstrous

Figure 14.4 Dickinson's two cat metaphors.

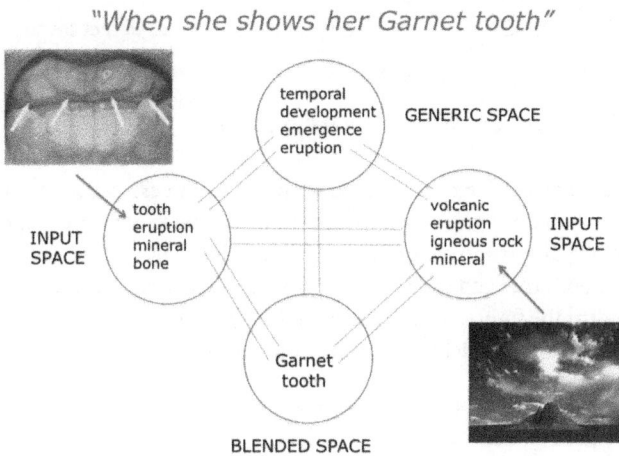

Figure 14.5 "Garnet tooth" blend.

beast roaring and spewing out from its mouth molten rock torn from the depths. I have marked relevant terms in **boldface**:

> A spreading bay is there, impregnable
> To all invading storms; and **Aetna's throat**
> With **roar** of frightful ruin thunders nigh.
> Now to the realm of light it lifts a cloud
> Of pitch-black, whirling smoke, and fiery dust,
> Shooting out globes of flame, with **monster tongues**
> That **lick** the stars; now huge crags of itself,

> Out of the bowels of the mountain torn,
> Its **maw disgorges**, while the molten rock
> Rolls **screaming** skyward; from the nether deep
> The fathomless abyss makes ebb and flow. (Book 3,
> lines 570–80)[3]

The dominating comparison in the poem is not between the two cat metaphors but between fear and security. The cat metaphors reveal Etna's increasing danger through time. One is fearful of a future unknown: what might happen, not of what does (Figure 14.6).

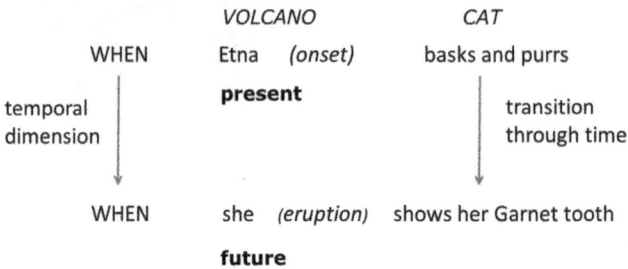

Figure 14.6 | Content |
| --- |

	VOLCANO		*CAT*
WHEN	Etna	*(onset)*	basks and purrs
	present		
temporal dimension			transition through time
WHEN	she	*(eruption)*	shows her Garnet tooth
	future		

Figure 14.6 The distance of fear: temporal dimension.

The more distant the danger, the more we imagine and exaggerate our fears. Both fear and security have external and internal components that are opposed, both positively and negatively, and connoting emotions of best- or worse-case scenarios. We can experience both true and false fear, true and false security. Fear is an emotion that is directed toward the future. Security is something we feel in the present. Dickinson's (Webster 1844) dictionary includes this definition for security:

> Freedom from fear or apprehension; confidence of safety; whence, negligence in providing means of defense. Security is dangerous, for it exposes men to attack when unprepared.

The idea of security as dangerous suggests that Dickinson is upending the conventional notions that fear is bad, security is good. Fear is represented on a spatio-temporal plane by

---

[3]   Portus ab accessu ventorum immotus et ingens   570
       ipse: sed horrificis iuxta tonat Aetna ruinis,
       interdumque atram prorumpit ad aethera nubem
       turbine fumantem piceo et candente favilla,
       attollitque globos flammarum et sidera lambit;
       interdum scopulos avulsaque viscera montis   575
       erigit eructans, liquefactaque saxa sub auras
       cum gemitu glomerat fundoque exaestuat imo.
       fama est Enceladi semustum fulmine corpus
       urgeri mole hac, ingentemque insuper Aetnam
       impositam ruptis flammam exspirare caminis,   580

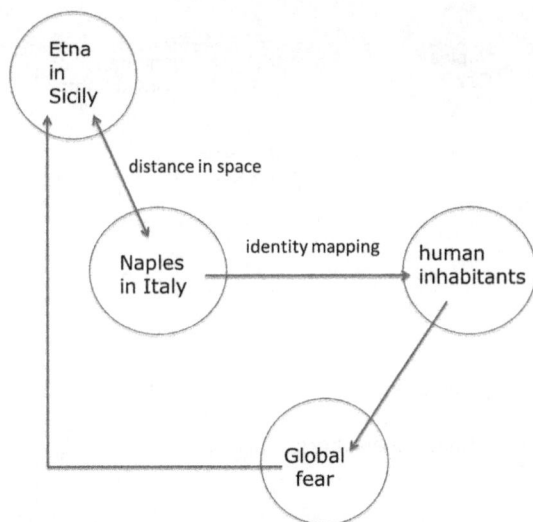

Figure 14.7 The distance of fear: spatial dimension.

"external" fears—those that are forced on the body from something outside—and "internal fears"—those that are part of the body even before being confronted with a dangerous situation. The contrast between the muted purr of the basking cat and the roaring sound of the wild cat mirror the distinction between fear of future possible volcanic activity versus the experience of the loudness of an actual eruption. The experience of past volcanic activity triggers fear of future danger in anticipating the worst that can happen.

Many critics, readers, and translators of this poem assume Dickinson erred in her geography by referring to Naples in relation to Etna. This is simply not true. Dickinson knew her geography. Located in the Gulf of Naples, an even more dangerous volcano—Vesuvius—suggests that Etna is a metaphor for any volcanic activity and for the notion of fear itself. The term *Naples* is a metaphor for the city that borders Vesuvius and for its human inhabitants as well as anyone physically distant from but emotionally affected by a volcanic eruption (Figure 14.7).

By invoking the distance of Naples from Etna, its human inhabitants experience a more global fear than their own familiar experience of Vesuvius. Fear in the poem becomes a metaphor for the internal feelings of apprehension, rather than the emotion of confronting danger (Figure 14.8).

In the end, there is a kind of security in knowing the worst. Dickinson expresses a similar attitude in other poems: "At least, to know the worst, is sweet!" (H15, F170/J172/M98) and "To know the worst, leaves no dread more -" (H54, F341/J281/M179).[4]

---

[4] I am grateful to Kang Yanbin for pointing out the Dickinson references. She notes that the poet may have been influenced by Emerson's (1941) comment in *Circles* that "'Blessed be nothing' and 'The worse things are, the better they are' are proverbs which express the transcendentalism of common life" <https://emersoncentral.com/texts/essays-first-series/circles/>).

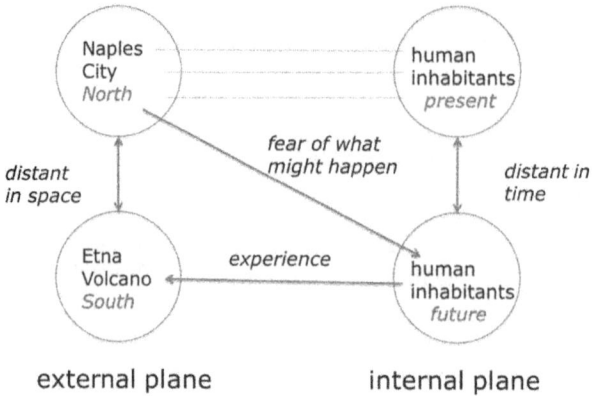

Figure 14.8 Spatio-temporal metaphorical plane.

Fearing the worst that can happen is imagining the possibilities to their utmost extreme. Dickinson's poem ironically reaches into our innermost feelings as we imaginatively envision the fear of watching the otherwise "safe" behavior of an incipient volcano gradually coming to life as it moves toward the greater menace of its eruption.

From a conceptual metaphor perspective, "basks and purrs" and "shows her Garnet Tooth" are not in opposition but are temporal stages of volcanic activity. By metaphorically mapping the actions of an active volcano onto those of a cat, Dickinson compresses our experience of a volcanic eruption, either physically embodied or mentally conceived, with the more intimate human scale of relating to another animal.

With its multiple layers of metaphoring, Dickinson's poem creates a scenario in which we as readers respond by exercising our own sensory-motor-emotive reasoning capabilities. At the surface is the mapping of a feline onto the volcano: "When Etna / basks and purrs." The image is intimate—one of a domestic cat soaking up the warmth of the sun and making a low, continuous, vibratory sound of pleasure. The introductory conjunction "when" sets up a temporal frame which, with its repetition in "when she / shows her Garnet tooth" indicates movement through time. This sense of movement is echoed by the transformation, in the next layer of metaphoring, of the domestic cat into the ferocity of a wild animal baring its bloody teeth. We are therefore led to understand the VOLCANO IS A CAT metaphor as one in which Etna also changes, transforming from its first signs of life into the violence of its eruption, from warmth and low vibration into the fiery heat and loudness of its explosion. Another metaphorical layer introduces the site of an even more dangerous volcano—Vesuvius—in the Gulf of Naples, suggesting that Etna may be a metaphor for volcanic activity in general.[5] The poem's reference

---

[5] In Dickinson's lifetime Vesuvius erupted seven times: 1834, 1839, 1850, 1855, 1861, 1868, 1872. Its neighbor Etna, to the south in Sicily, even larger and more active, erupted six times: 1832, 1852–3 (lasting ten months), 1863, 1865, 1879, 1883. Of the two, Vesuvius is considered the more dangerous.

to "Naples is more afraid" may be taken as a metaphor for the city and its inhabitants as well as anyone physically distant from but emotionally affected by the signs of an imminent (and immanent) volcanic eruption.[6]

The poem is thus the result of complex blending of several metaphorical layers through compression to human scale. The poem's last line, "Security is loud" is a typical Dickinsonian ending which, in Alice Parker's terms, "turns her thought right at the end to make an unexpected conclusion" (personal correspondence). It is this line that creates a puzzling complexity to the poem and calls for an even closer examination of the cognitive processing at work.

Understanding *cognitive* as meaning *conceptual* is a holdover from the traditional body-mind split of Cartesian thought that separated reason and logic from sensory perception and emotion. Although attention to the integration of sensate cognition and reasoning is evident among philosophers and others over the centuries since Descartes, like Giambattista Vico (1948[1744]) in the eighteenth century and George Henry Lewes (1879) in the nineteenth, it is only with the work of the cognitive sciences in the latter part of the twentieth century, especially by cognitive psychologists and neuroscientists, that *cognitive* has more generally been embraced to include the role of the subliminal sensate elements as integral to human minding (Damasio 1994, 1999; Le Doux 1996; Forgas 2000; among many others).

Metaphoring occurs at an even deeper level in the poem with the seemingly paradoxical notion that "Naples is more afraid" of Etna's "basks and purrs" than seeing Etna's "Garnet tooth." One is naturally led to ask why one would be afraid of the seemingly innocuous behavior of VOLCANO AS DOMESTIC CAT basking and purring and less afraid of VOLCANO AS WILD FELINE showing her garnet tooth. As Liane Ströbel (2014, 150) notes, "a higher categorization level of the underlying source domains of the fear metaphors becomes distinguishable" between "internal fears"—those that are part of the body even before being confronted with a dangerous situation—and "external" fears—those that are forced on the body from something outside.

Fear is an emotion that is directed toward the future. That is, one is fearful of what might happen, not of what does. One is not fearful when physically confronted with danger; rather, one might be scared, terrified, frightened, or any other relevant emotion. In other words, fear involves anticipation or apprehension of something bad happening that has not yet happened. At the deepest layer of metaphoring then, fear in the poem becomes a metaphor for the internal feelings of apprehension, rather than the emotion of confronting danger. It is a characteristic Dickinsonian position, of focusing on anticipation and aftermath to the exclusion of present experience (M. Freeman 2005a; D. Porter 1981). It is no surprise, therefore, that the poem ends with "Security is loud." The metaphor links "loud" with "Garnet tooth" and an actual eruption as opposed to the "basks and purrs" of the not yet exploding volcano, at the

---

[6] "Imminent" means something about to happen in the near future; "immanent" is synonymous with "inherent" and can be defined as "being within the limits of possible experience or knowledge."

same time as it identifies "Security" as the opposite of fear. Dickinson's (Webster 1844) dictionary has the following entry for fear:

> A painful emotion or passion excited by an expectation of evil, or the apprehension of impending danger. Fear expresses less apprehension than dread, and dread less than terror and fright. The force of this passion, beginning with the most moderate degree, may be thus expressed, fear, dread, terror, fright. Fear is accompanied with a desire to avoid or ward off the expected evil. Fear is an uneasiness of mind, upon the thought of future evil likely to befall us. Watts. Fear is the passion of our nature which excites us to provide for our security, on the approach of evil. Rogers.

After all, there is a kind of security in knowing the worst. Fearing the worst that can happen is imagining the possibilities to their utmost extreme. Torturers throughout history have employed anticipatory tactics to heighten fear (Burman 1984, 146–50). Dickinson's poem ironically reaches into our innermost feelings as we imaginatively envision the fear of watching the otherwise "safe" behavior of the incipient volcano as it gradually comes to life to move toward the greater menace of its eruption. The poem thus becomes an icon of the very notion of fear.

5 The Transformative Power of the Poetic Icon

Dickinson's poetics of CONTAINMENT and TRANSFORMATION are realized in the following poem whereby the very form of the poem creates the poem as an icon that reaches toward that which is beyond itself:

> My Cocoon tightens -
> Colors tease -
> I'm feeling for the Air -
> A Dim Capacity for
> Wings
> + Demeans the Dress I wear -   +
> Degrades
>
> A power of Butterfly must
> be -
> The Aptitude to fly
> Meadows of Majesty
> + concedes   + implies -
> And easy Sweeps of Sky -
>
> So I must baffle at
> the Hint

> And Cipher at the Sign
> And make much blunder
> if at last
> I take the Clue divine -

H189b, F1107/J 1099/M494

From a traditional perspective, the metaphor of the poem seems clear enough. The speaker or persona of the poem is likening herself to a butterfly about to emerge from the container of its cocoon.[7] Literary critics interpret the poem by inferring the cause and purpose of this mapping. For instance, from a feminist perspective, the butterfly may be seen as a woman resisting patriarchal restriction. From a metaliterary point of view, it may represent the power of poetry breaking free from the constraints of prose. From a theological perspective, the butterfly may be a sign of the Resurrection. All these readings are possible. They do not, however, explain what enables the critic to draw such analogies. They are possible because they are all coherent within the poem's cognitive scope.

Dickinson's three-stanza format lends itself to the possibility of the CONTAINMENT schema, realized here by several relations existing between the first and last stanzas. Unlike the central stanza, both are voiced by an "I" repeated twice in each stanza:

> I'm feeling / I wear
> I baffle / I take.

The two stanzas are related by mappings from the concrete to the abstract. The concrete images of the first stanza—cocoon, colors, wings, dress—are mapped onto the abstract terminology of the last—hint, sign, cipher, clue. As these mappings move from the concrete to the abstract, they work on both structural and semantic dimensions.

In the poem as a whole, with attribute mapping, the source domain of the butterfly is mapped onto the target domain of the speaker, the cocoon onto dress, and the development ("dim capacity") of wings onto the speaker wanting to break out. At the relational level, the tightening of the cocoon is mapped onto the speaker's feeling the constraint of enclosure, the teasing of colors onto the attraction of becoming something different, and the "dim capacity for wings" onto the idea that the something that wants to break out is better and more powerful (since it "demeans" or "degrades") than the physical, emotional, or mental constraints binding the person.

---

[7] Although entomologists distinguish between the words *cocoon* and *chrysalis* in the life stages of butterflies and moths, it seems that Dickinson used the words interchangeably (see the Emily Dickinson Lexicon at edl.byu for details). In one poem, *Cocoon above* (H10, F142/J129/M80), she combines both words in her description of the process of an emerging butterfly escaping imprisonment.

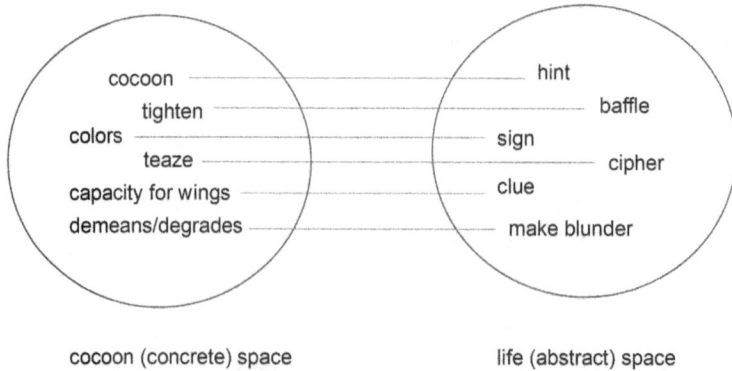

Figure 14.9 Identity mapping between the natural and the human.

Syntactically, the three verbs and their subjects in the first stanza are mapped onto the three verbs and their complements in the last. Thus, "tighten" maps to "baffle" and "cocoon" maps to "hint," "tease" to "cipher" and "colors" to "sign," "demean" (or "degrade") to "blunder" (since to make blunder is to blunder), and "capacity" to "clue." The "'cocoon," in its temporary state, "hints" at its metamorphosis; "colors" are the "sign" of the emergent butterfly; and the "dim capacity for wings" provide a "clue" toward transcendence (Figure 14.9).

On the semantic level, the verb and noun phrases metaphorically map onto each other, as the poem dynamically progresses from the effect, in the first stanza, of the physical environment represented by the cocoon to the response, in the last, of the human persona to the environment of life. In the process, the focus changes from the constraint of the cocoon itself to the persona struggling within that constraint, and the verbs reflect the transformation. Thus, "tighten" transforms to "baffle," as the speaker struggles ineffectually (the intransitive meaning of the word) with the "hint" of the "cocoon"; "tease" transforms to "cipher," as the sensual, physical idea of "colors" appearing briefly as a glimpse, then hiding, transforms to the intellectual puzzling at the glimpse of the "sign"; and "demean" or "degrade," caused by the "dim capacity for wings," transforms to making a "blunder" as the persona moves blindly in groping toward the "clue divine."[8]

System mapping connects the semantics of cocooning with the structure of the poem. In drawing the imaginary, overlapping lines that relate the images of the source domain (butterfly) in the first stanza to the abstract terminology of the target domain (person) in the last stanza, a container—a cocoon—has been created around the stanza in the center (Figure 14.10).

Whereas the outer stanzas thus together manifest the CONTAINMENT schema, the central stanza manifests Dickinson's other schema of TRANSFORMATION. The "I" of

---

[8] The word *degrade* has a specific biological meaning of "to reduce to a lower and less complex organic type" (*OED*).

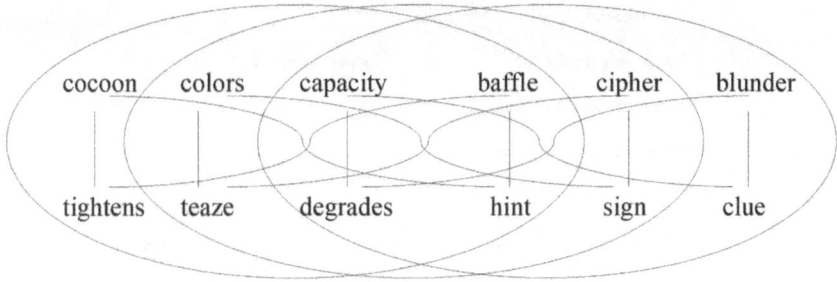

Figure 14.10 Cocoon image.

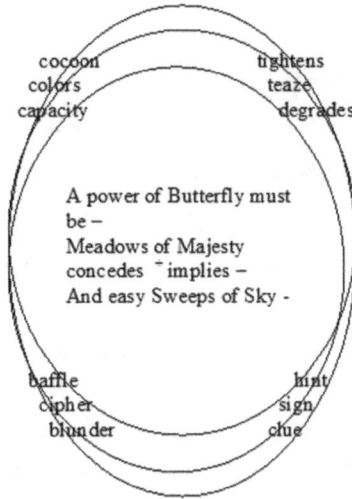

Figure 14.11 Iconic realization of the CONTAINMENT and TRANSFORMATION schemas.

the first stanza is that of the pupa within its cocoon. The "I" of the last stanza is that of a human cognitively searching. In the central stanza the voice changes to neither, but rather portrays a meditation on the power of freedom. And what is in that stanza but the potentiality of the butterfly itself? Not the caterpillar in its prior state or the pupa within its chrysalis, but the fully metamorphosed form of the butterfly to be, potentially freed from the container of its cocoon to fly in its full power of majesty over meadows and through "easy Sweeps of Sky." The first and last stanzas of the poem surrounding the central stanza thus iconically form a cocoon for the future butterfly it contains. Peirce's image, diagram, and metaphor thus all cohere in the poem's iconicity (Figure 14.11).

From a cognitive mapping perspective, however, the poem is incomplete. It is "missing" a fourth stanza, in which, as stanza 1 maps to stanza 3, stanza 2 should map to stanza 4. That is, the cocooning stage (insect: literal) of stanza 1 maps to the

My Cocoon tightens - Colors teaze -

I'm feeling for the Air -   Cocoon: Insect

A dim capacity for Wings

⌐Demeans the Dress I wear -

A power of Butterfly must be -

The Aptitude to fly   Butterfly: Insect

Meadows of Majesty ⌐concedes

And easy Sweeps of Sky -

So I must baffle at the Hint

And cipher at the Sign   Cocoon: Human

And make much blunder - if at last

I take the clue divine -

+ Degrades   + implies -

                    ?   Butterfly: Human

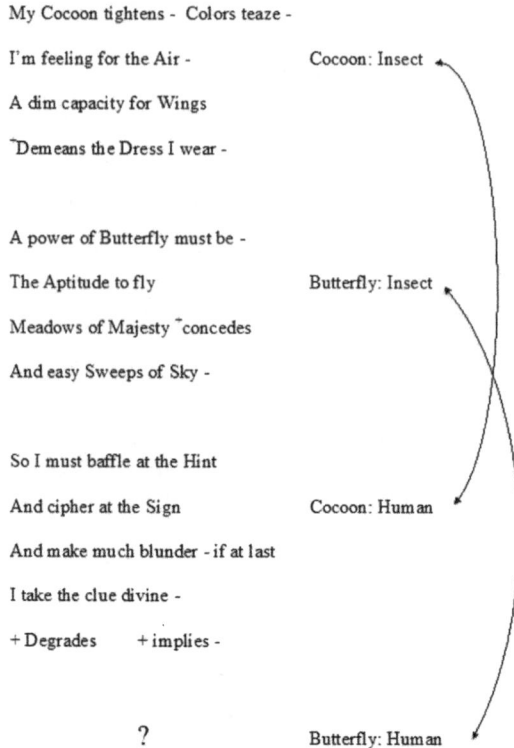

Figure 14.12 The missing stanza.

cocooning stage (human: metaphor) of stanza 3; the butterfly stage (insect: literal) of stanza 2 should map to the butterfly stage (human: metaphor) of the missing stanza 4 (Figure 14.12).

From the perspective of poetic cognition, the theory predicts that there should be a fourth stanza and explains what would be there if there were one. It shows why the poem works elegantly in not having a fourth stanza and demonstrates it as an artwork about art reaching beyond the merely representational. The poem as it stands exists at the cocoon stage, the potentiality of the butterfly still contained within the frame of the outermost stanzas, its human analog only suggested, not presented. The beauty of Dickinson's poem lies in its incompleteness, the suggestion left in the reader's mind's eye of the "Clue divine" that ends the poem: human life existing in potentiality, not actuality, human aspiration, and poetry as possibility.

Given the isomorphism created by these structural mappings, we understand the poem according to the purpose and cause of the analogical mapping, as has been noted: in terms of a woman resisting restriction, the power of poetry, or a symbol of the Resurrection. These readings (and they are only a few of the possible ones) are consistent with the analogical mapping of the prototypical isomorphic metaphor arising from Dickinson's schemas of CONTAINMENT and TRANSFORMATION.

It is the power of our cognitive capacities for memory, imagination, attention, discrimination, expertise, and judgment that exist at the heart of the aesthetic faculty, thus enabling us to probe beyond the physical world to bring absence into presence. In Dickinson's succinct summation:

> The gleam of an heroic act
> What strange illumination
> The Possible's slow fuse is lit
> By the Imagination

HST Ia, F1686/J1687/M662

For Dickinson, then, the potentiality of poetry lies within its iconic reaching out beyond the limits of human experience into the realms of the imagination as we attempt to understand the hows and the whys of creation.

Dickinsonian Cognition

The previous chapters have focused on the various ways a cognitive approach can illuminate a reader's experience of Dickinson's poetry and thus draw the reader more closely to an understanding of the how and the why of what a poem is saying and doing. A cognitive approach does so by:

1. recognizing the ways in which creative metaphor works;
2. identifying the underlying schemas that link the bodily subliminal processes of sensory, motor, and emotive faculties to conceptual awareness; and
3. exploring the various element of prosody that reveal a poem's affective force.

A cognitive approach therefore does not simply offer yet another possible interpretation to a poem among the many already proposed. It makes explicit how and why a poem as perceived by the reader is doing what it is doing in simulating human experience. Rather than providing yet another literary interpretation of a poet's œuvre, reading cognitively both opens up and constrains the possibilities of a poetic text. It thus provides an effective test for literary experience and interpretation.

In this conclusion, I go further to (1) show how cognitive analysis can identify and authenticate a poet's poetics and (2) suggest an explanation for poetic evaluation: what makes a poem succeed in what the poet set out to do and perhaps provide the reason for its endurance through time and across cultures.

What makes a Dickinson poem recognizable as such? The brief answer is that no one feature, whether it is handwriting, markings, word choices, etc., can do so, but all of them contribute, especially the imagistic and diagrammatical relationships that cohere within the poem taken as a whole. As I. A. Richards wisely noted in the era of New Criticism, any literary theory should be able to recognize literary style, to be able to distinguish the characteristics of an individual writer. He also noted ruefully that New Criticism failed this test when his students were unable to identify beyond any doubt the author of a text he gave them. Even today, current literary theories are unable adequately to identify poetic style, as evidenced by the failure of many literary critics to determine that a poem claimed to be by Dickinson was in fact a forgery. In what follows, I discuss the ways in which a cognitive approach shows that Dickinson could not have written the poem attributed to her.

1 *The Jones* Manuscript: A Dickinson Forgery

Published in 2002, Simon Worrall's book, *The Poet and the Murderer: A True Story of Literary Crime and the Art of Forgery*, focuses on the story surrounding the forgery of a Dickinson poem, Dickinsonian Brent Ashworth's connection to the affair, the murders the forger Mark Hofmann committed, and the details of how Dan Lombardo, Special Collections Curator at The Jones Library in Amherst, subsequently uncovered the manuscript's provenance as a Hofmann forgery. It is a lively tale for those wishing to read it. The book, however, does not tell the true story of how Dickinsonians came to recognize the poem as a forgery. That is the story I tell here.[1]

It has been almost a quarter of a century since the summer events of 1997. On Tuesday, June 3, 1997, with the help and generosity of members of the Emily Dickinson International Society (EDIS) and other Dickinsonians, Lombardo successfully bid at a Sotheby's auction on Lot #74 for an Emily Dickinson manuscript that contained the text of a poem previously unpublished. Seth Rothenburg transcribed the poem from the photograph in the Sotheby catalog and circulated it online among Dickinsonians:

> *Unpublished Dickinson Poem*
> That God cannot
> be understood
> Everyone Agrees
> We do not know
> His motives nor
> Comprehend his
> Deeds
>
> Then why should I
> Seek solace in
> What I cannot
> Know?
> Better to pray
> In winter's sun
> Than to [fear] the
> Snow.[2]

When I read Rothenburg's transcription, I immediately thought it was not Dickinson. So I set up a forum on the emweb listserv to see what others thought. The manuscript had been examined by several Dickinson scholars and its authentication certified by

---

[1] A full account can be found at: https://www.academia.edu/44991675/The_Jones_Manuscript_A _Dickinson_Forgery.

[2] Rothenburg took his transcription from the copy in Sotheby's catalog. He removed the markings since they were unclear, and bracketed the word *fear* because he was unsure of its reading. Georgie Strickland pointed out that "pray" should read "play."

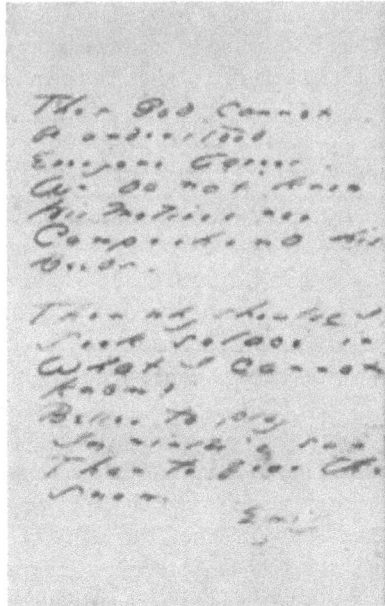

Figure 15.1 The Hofmann Manuscript Courtesy Catherine Eisner (photo copy from Sotheby's catalog).

Ralph W. Franklin, the editor of Dickinson's manuscripts. Discussion thus centered around the poem itself and the fact that the handwriting in the manuscript had been established as Dickinson's. Once established as a forgery, it was obvious that Dickinson had not and could not have written the poem. However, it is understandable that Dickinsonians were divided on the question, given its certification by experts, and several arguments were put forward on its behalf.

When Lombardo finally received the actual document he had successfully bid on, he sent me a facsimile. For several years, I had been working in Los Angeles with Susanne Shapiro, a professional graphologist, on Dickinson's handwriting.[3] Graphologists are frequently called upon in court cases to confirm authenticity of signatures. As soon as I saw a copy of the original, I knew it was not Dickinson's handwriting. I sent it to Shapiro who confirmed my analysis. The formation of the letters is carefully deliberate, without the flowing of fictive motion that handwriting automatically has, even when letters are separated within a word (Figure 15.1).

That clinched the matter for me, but the story of its provenance was still not known. Lombardo, made uneasy by the possibility that the manuscript was not Dickinson's, had begun his enquiries into its provenance, while I began to wonder why it was that I was so sure that the poem was a forgery, even before seeing the manuscript, when

---

[3] Shapiro (2001) presented her findings at the third International Emily Dickinson Society conference at Mount Holyoke in 1999, subsequently published in the conference proceedings.

eminent Dickinsonian experts accepted it as genuine.[4] I could only conclude that it resulted from the work I had been doing in cognitive approaches to literature. And at that point, I had only my judgment that neither the poem itself nor the manuscript were authentic.

2 The Limits of Forgery: A Question of Cognitive Style

Dickinson's characteristic stance in her poetry is to overturn the stereotypical and superficial metaphors by which we construct our folk theory of the world. What happens in the "Dickinson" poem Mark Hofmann forged?

Ignoring other elements of poetic style, where in this poem is the disruption of the commonsensical and the stereotypical? The opening lines assert the very opposite of a Dickinsonian stance: the word "agree" occurs nowhere in Dickinson's poetic lexicon: her usual stance is to disagree, not to conform. The second stanza, moving from the "we" of the first to the "I" of the second, is a version challenging Pascal's wager that it is better to act as if we believed in God than not. The amorality expressed in Hofmann's lines is antithetical to Dickinson's thought processes. The final sentence rings false. Dickinson would never have succumbed to such a platitude. In fact, the whole poem is platitudinous. We do know that Dickinson sometimes expressed admiration for the platitudinous, but she was incapable—at any stage of her life—in producing it.

When I first read the text in Seth Rothenburg's transcription, I memorized it within seconds. This was my first inkling that it wasn't a poem Dickinson would have written. I had spent several months working on a fourteen-line manuscript poem (six in the edited version) and still stumbled over speaking it aloud from memory. Dickinson's complicated syntax and seemingly irregular punctuation and pauses make it difficult to remember lines accurately. The reason the Hofmann poem was so easy to memorize was its regular metrical rhythm and its straightforward syntax. Its rhythmic regularity was overly metronomic. The underlying structure of English syntax is subject-verb-object (SVO). As we have seen, Dickinson often preposes the object to create variations of OSV and SOV which other languages like Hebrew and German have as their underlying structures. A simple example can be seen in the last two lines of the following poem that exists in transcript only. Both lines and their variant would conventionally be written as: "that we would rather play with each other than with him."

> God is indeed a jealous God -
> He cannot bear to see
> That we had rather not with Him
> But with each other play.

---

[4] Polly Longsworth subsequently informed me that even before the auction, Lombardo had read the text of the poem over the phone to her and she had told him that "it wasn't Emily's poem" (personal correspondence). So he must have already been made uneasy.

> [3–4] That we desire with ourselves
> And not with Him to play.
>   (A Tr10, F1752/J1719/M680; no autograph copy)

Even without the original manuscript, the poem is definitely Dickinson's. She is setting up two mental spaces in this poem within the cognitive domain of "play": that is, to play with someone means to be physically in their presence (this before the age of telephonic and electronic capabilities of playing at a distance). Thus, to play with each other means being in the jointly experienced physical space of life; to play with God means to move into his space, which for us is death, since we "cannot know God and live" to quote the Old Testament. The poem thus plays with the two domains of life and death. In *The Jones* manuscript, although a similar comparison is set up—better to play in the sun of winter than to fear the snow—there is no crossing of domain spaces. The contrasting mental spaces of sun and snow in the forged poem occur within the same cognitive domain of winter, they do not cross domains like the life-death domains of "God is indeed a jealous God." For a poet who agonized over the snow of death and the very question she said disquieted her—what, if anything, survives after death, the forged poem's tone of certitude—"cannot be understood," "we do not know," "we cannot know"—does not ring true.

Other characteristic Dickinsonian formulations—her frequent ellipses of words and phrases, her idiosyncratic uses of pronouns, and many other grammatical forms—are nowhere to be found in the Hofmann poem. Dickinson disrupts the stereotypical through complex inverted syntax and the creation of multiple mental spaces; none exist here. There is no complex metaphorical reversal of figure and ground; the deictic perspective is unrelentingly "ours." The poem ends with the reinforcement of the conventional— the choice of sun over snow. Worst of all is the platitudinous litany of our ignorance in the face of an almighty God and the consequent assertion of an amoral hedonism. Dickinson must have been shuddering in her grave when her readers ascribed this poem to her. Compare the forged poem with the prospectives set up in a poem of similar length which exists in transcript only, no manuscript having yet been found:

> That it will never come again
> Is what makes life so sweet.
> Believing what we don't believe
> Does not exhilarate.
> That if it be, it be at best
> An ablative estate -
> This instigates an appetite
> Precisely opposite.
>   (A Tr28, F1761/J1741/M682; no autograph copy)

The negative reversals in this poem are typically Dickinsonian in their complexity, especially in the second sentence, where the oxymoronic "Believing what we don't believe" makes me pause until I see how it cannot "exhilarate." The following hypothetical mental space "if it be [believed]" describes what has been omitted in the previous sentence as "an ablative estate," thus reversing the conventional desire for the afterlife to "an appetite / Precisely opposite."

Several Dickinsonians in the online discussion forum identified liftings from other poems such as the phrase "fear the snow" in *Distrustful of the gentian* (A82-1/2, F26/J20/M34). In this poem it is the autumn gentian that encourages the speaker to face the snows of winter, thus revealing a mental space crossing of seasonal domains. Another identification in the use of the word *solace* in "Why should I seek solace" does not characterize the stance Dickinson takes in other poems. In the lines "Some have solaced the longing" from *Not any more to be lacked* (H380, F1382/J1344/M522), and "At least, it solaces to know" from *Your riches taught me poverty* (H47, F418/J299/M165), the speaker indeed considers the possibilities of solace.

A review of corresponding words that Dickinson uses turns up three other poems that Hofmann may have had in mind in creating his forgery (corresponding words in boldface):

Back from the Cordial Grave I drag thee
He shall not take thy Hand
Nor put his spacious Arm around thee
That **none can understand**

A144, F1649/J1625/M650

His Mind like Fabrics of the East -
Displayed to the despair
Of **everyone** but here and there
A humble Purchaser -
For though his price was not of Gold -
More arduous there is -
That one should **comprehend** the worth -
Was all the price there was -

A211,211a, F1471/J1446/M613

Those - dying then -
Knew where they went -
They went to God's Right Hand -
That Hand is amputated now
And **God cannot be found** -

The abdication of Belief
Makes the Behavior small -
Better an ignis fatuus
Than no illume at all -

A476, F1581/J1551/M638

In these cases, Hofmann's usage has a directly opposite effect from Dickinson's. This is also true of the word *deeds*. All eight instances in the poetry refer to people or nature's deeds. God doesn't have deeds. In fact, the following poem contemplates the response of God to a deed that is merely contemplated:

> A Deed knocks
> first at Thought
> And then - it knocks
> at Will -
> That is the manu-
> factoring spot
> And Will at Home
> and well
>
> It then goes out an
> Act
> Or is entombed so still
> That only to the Ear
> of God
> It's Doom is audible -

A98, F1294/J1216/M570

Dickinson's complications of syntax reflect the complications of her content. Dickinson is a complex thinker, one who questioned and argued and upended clichéd and stereotypical attitudes. The God of Hofmann's God is not Dickinson's God. Dickinson had certain fixed and varied attitudes about "her" God: a God who denied Moses access to the Promised Land, a God who went around with a telescope, a God who was a "stately - distant Lover"; just a quick look at the 130 uses of the word *God* in Rosenbaum's (1964) *Concordance* indicates how diverse her feelings about God are. Nor is the casual easy statement of "everyone" Dickinsonian, or the word *agrees* that is found nowhere in the entire extant corpus of her writings, in poems or letters. It was, however, the second stanza that confirmed for me the poem as a forgery. It is so obviously anti-Dickinsonian, just as Hofmann's forgeries of Joseph Smith's writings were (deliberately) anti-Mormon beliefs.[5]

In Dickinson's poem, *God is indeed a jealous God*, the lack of understanding is not ours, it is God who does not understand. It is just this audacity—to jump into God's perspective and then to undermine it—that is characteristically Dickinsonian. The banality of the forged poem should in itself have raised the suspicions of Dickinson scholars.

---

[5] As I understand it, Hofmann was astute in preparing his forgeries for easy sale. He knew that not only would the Church Elders want copies of Smith's original manuscripts, they would also want even more any copies that seemed to be in violation of the Church's mandates. In fact he was right. The Elders apparently bought the manuscripts from him so they could hide them in the Vaults to prevent their congregations from being led astray. That is yet another dimension to this very complex story. I am grateful to Dr. Robert G. Kent de Grey for identifying my source for this information from Jerald and Sandra Tanner of Utah Lighthouse Ministry.

But this is just my point. It is not that Dickinson scholars are not sensitive readers or accomplished critics; they are. Poetic cognition provides a very clear means by which we can read and describe and identify a literary text and its style.

3 Aesthetic Evaluation

A thorny question in literary criticism is that of evaluation. Who is to say whether a poem succeeds in what the poet set out to do? What is it that connoisseurs of the arts know that enables them to pronounce a work of art mediocre, good, or great? One criterion is that it endures in speaking across centuries and cultures. But that doesn't explain why it does so. To speak of aesthetics as simply taste, beauty, and pleasure in the arts is neither sufficient nor accurate, since these are products, not functions, of the aesthetic faculty (M. Freeman 2020, 141–3). The aesthetic faculty involves purpose, intension, function, and value ascribed to our imaginative understanding. It reflects the ability to discern, to discriminate, thus to distinguish qualitatively the successful from the unsuccessful, good from bad, right from wrong, the ugly from the beautiful, the true from the false. The aesthetic faculty underlies all human activities, including the natural and social sciences as well as the arts and humanities. It is not innate. It is acquired through education and intelligent design that leads to the capabilities of purpose, discrimination, expertise, judgment, and creative activity.

It is beyond the scope of this book to develop a full theory of aesthetic evaluation in identifying the crucial role the arts play in human cognition. However, Dickinson's poetry leads to some suggestive indications of what characteristics might be included in such a theory: the way cognitive schemas and metaphor can structure a poem; the repetitive interplay of sounds and rhythms that invoke affective response in the reader; the way words are made to work; and, above all, a transformative reaching out to what is always beyond human grasp in creating a poem as icon.

In 1877, on the third anniversary of Dickinson's father's death, the poet wrote to Higginson expressing her doubt about achieving immortality (L503). In it, she wrote of re-reading his "Decoration," a poem of seven quatrains and enclosed a quatrain of her own.[6] Dickinson's poem reads as follows:

> Lay this
> Laurel on
> The One
> Too intrinsic
> for Renown -
> Laurel - vail
> your deathless

---

[6] Higginson's poem, written to commemorate Newport, Rhode Island's Decoration Day in May 1973, was published in Scribner's in 1974. It may be found at https://www.poemhunter.com/poem/decoration-2/. A full description of the exchange is given in T. H. Johnson (1963, 960–2). See also Mary Loeffleholz's (2005) "Dickinson's 'Decoration.'"

> Tree -
> Him you
> chasten ⸝ That
> is He!

A268/A269, F1428AB/J1393/M600)

Higginson later wrote Mabel Loomis Todd, saying that Dickinson "wrote it after re-reading my 'Decoration.' It is the condensed essence of that & so far finer" (Bingham, 128–30). The question is: What is it about Dickinson's poem that makes it better than Higginson's?

Higginson's poem has its epigram the Latin *manibus date lilia plenis* (give lilies with full hands), a quote from Virgil's *Aeneid* VI.883, mourning the death of Marcellus, Augustus's nephew, and subsequently quoted in Dantes's *Purgatorio* XXX.21, as the messengers of life eternal strew lilies on Beatrice's arrival. His poem opens with the lines: "Mid the flower-wreathed tombs I stand / Bearing lilies in my hand," and asks the question "in what soldier-grave / Sleeps the bravest of the brave?" The second stanza asks a second question: "Is it he" who lies veiled by garland, then introduces what will be the subject of the following quatrains—the person in the grave that still lies unrecognized, unadorned—before the closing quatrain:

> Turning from my comrades' eyes,
> Kneeling where a woman lies,
> I strew lilies on the grave
> Of the bravest of the brave.

As Mary Loffelholz (2005) notes, several commentators missed Higginson's point of celebrating the truly unknown, the women who played their part in the Civil War, as being "the bravest of the brave." I suggest there are two reasons why his point was missed: the mention of a "soldier-grave" in the opening question and the ambiguity in stanzas 3–6 that do not define gender and therefore may lead to the assumption, subsequently false, that it is an unknown male soldier that is being referenced.

Dickinson's compression of Higginson's theme into one stanza makes Higginson's point, otherwise lost, more succinct. But it is doing a lot more than that. Each word in the poem is being made to work. First, Dickinson cuts to the chase by immediately changing Higginson's lilies to the Laurel—the leaves of the Bay tree—that comprise the garland of honor. Second, Dickinson changes the question format of Higginson's poem into an invocation—"Lay this Laurel"—thus drawing the reader directly into participation.

In turning to address the Laurel, Dickinson engages in a different kind of word-play. Whereas Higginson's poem has garlands veiling a tomb, Dickinson's poem includes a second invocation to the Laurel itself: "vail / your deathless / Tree," thus focusing on the idea of immortality, not death.[7] The syntax of the final two lines refers specifically

---

[7] In Christianity, the bay tree symbolizes Christ's resurrection (https://www.ancient-symbols.com/bay-tree).

to the Laurel's role of chastening—recognizing through purification—in such a way Higginson's question, "Is it he?" is answered by the final statement "That / is He!"

Higginson's poem is one of comparison: Who is the bravest of all who have died in war? It has the effect of pitting the particular achievements of one against others, instead of celebrating one who is "Too intrinsic / for Renown." That, for me, is the "eye" of Dickinson's poem. Webster gives the following definition of intrinsic: "Inward; internal; hence, true; genuine; real; essential; inherent; not apparent or accidental; as, the intrinsic value of gold or silver; the intrinsic merit of an action; the intrinsic worth or goodness of a person." By placing the word *intrinsic* in comparison to *renown*, Dickinson's comparison, unlike Higginson's, points to that which is beyond renown, into the realm of Merleau-Ponty's (1968) *in-visible*, the realm of the icon.[8]

4 Poetic Cognition

Why should poetic cognition emerge as an important theory now? With increasing neuroscientific research into the workings of the human brain, developments in cognitive psychology, and emerging theories of embodied cognition, we are discovering more about how we think and feel about ourselves and the world of which we are a part. The capabilities of human cognitive processes as they relate to artistic creation become an important component in discovering the many ways we respond to and express those thoughts and feelings. Poetic cognition can provide a principled explanation for artistic production and motivation. It starts with language and not with ideology; it includes affective cognitive processes together with the contextual/cultural dimensions of what Jordan Zlatev (1997) has called "situated embodiment"; it can account for the capability of a literary text to generate multiple meanings; it can explain what it is literary critics do when they interpret a literary text. Most importantly, it can be tested.

As I have tried to show, there are at least four additional ways in which a cognitive approach can contribute to literary analysis:

1. it can explicate and possibly evaluate the differences among multiple readings;
2. it can identify what cultural knowledge influences interpretation;
3. it can contribute to a clarification of the writer's—as opposed to the reader's— conceptual worldview; and
4. it can form the basis for an empirical study of literary interpretations.

A cognitive approach integrates different literary approaches by showing how they make use of cognitive mapping strategies across cognitive-cultural domains in exploring the conceptual processes of the writer, reader, and text. Understanding the cognitive

---

[8] The in-visible is Merleau-Ponty's (1968, 257) primordial precategorial, that which exists before our conceptualizing brings experience into consciousness: "what exists only as tactile or kinesthetically, etc."

principles of poetic creativity can illuminate corresponding cognitive research into the way humans think and feel.

Words are not "dead when they are said" as Dickinson knew; they are constantly changing across time and space. Their vitality lies in the thoughts they stimulate. The following poem expresses Dickinson's aesthetic understanding of this truth:

> The Poets light but
> Lamps -
> Themselves - go out -
> The Wicks they
> Stimulate
> If vital Light
>
> Inhere as do the
> Suns -
> Each Age a Lens
> Disseminating their
> Circumference -

A91-13/14, F930/J883/M436

Meanings may change as each age applies its own lens, but Dickinson is saying something more in this poem. Its syntactic structure consists of three lines at the beginning and three lines at the end that frame the five lines in between: "The Poets light but / Lamps - / Themselves - go out - / . . . / Each Age a Lens / Disseminating their / Circumference." Lines 1–3 and lines 9–11 comprise together a coherent thought expressed metaphorically: Although poets die, future ages will keep their poetry alive. The five lines in the middle also divide into two parts that form a complete sentence: "The Wicks they / Stimulate / . . . / Inhere as do the / Suns," interrupted by the conditional clause "If vital Light." It is no accident that this line, "If vital Light," occurs in the exact center of the middle section and the exact center of the poem. It serves as central point, the "eye" of the poem. In typical Dickinsonian fashion, the word *vital* applies both backward to the word *Wicks* and forward to the word *Light*.

The analogy *as do the suns* suggests active agency on the part of the suns in making light inhere. This transitive meaning of the verb *inhere* is very rare. The word *inhere*, from Latin *in-* "in" plus *haerere* "to stick, remain fixed," suggests permanence. Such permanence of poetic creation is thus constrained by the conditional: "If the wicks poets stimulate [in lighting lamps] are vital, then light inheres in them as suns inhere light." The poem reverberates around the words *if vital light*. The first meaning of *vital* in the *OED* is as follows: "Consisting in, constituted by, that immaterial force or principle which is present in living beings or organisms and by which they are animated and their functions maintained." As we saw in the Genesis creation story, light is life (see Chapter 6). For Dickinson, words are vital when poets succeed in giving them life (H ST14c-d, F1715/J1651/M671):

> A word that breathes distinctly
> Has not the power to die
> Cohesive as the Spirit
> It may expire if He [does] -

The spirit is the life-force that animates a poem into being, that makes it breathe. The wicks poets stimulate are both the essence of the poem itself motivated by the poet and the inspiration kindled in readers as they respond. When that happens, Dickinson claims, a poem has achieved the iconicity of endurance through infinite space and eternal time.

What, then, is the conclusion we can draw about the aesthetics of poetry (and literature in general)? The September 2004 issue of *Harper's* magazine included an article by the novelist Tom Robbins called "In Defiance of Gravity," on the aesthetics of good writing. He tells the anecdote of the musician who barked on Big Mama Thornton's recording of "Hound Dog," and who is reported to have said "Yeah, . . . I was gonna meow—but it was too hip for 'em." Although, Robbins says, our culture "simply has a far greater demand for the predictable bow-wow than for the unexpected caterwaul," it is the cat's meow that marks the "crazy wisdom" of the poets:

> Crazy wisdom is, of course, the opposite of conventional wisdom. It is the wisdom that deliberately swims against the current in order to avoid being swept along in the numbing wake of bourgeois compromise; wisdom that flouts taboos in order to undermine their power; wisdom that evolves when one, while refusing to avert one's gaze from the sorrows and injustices of the world, insists on joy in spite of everything; wisdom that embraces risk and eschews security; wisdom that turns the tables on neurosis by lampooning it; the wisdom of those who neither seek authority nor willingly submit to it. (58)

Had Dickinson been ruled, had she organized, she wouldn't have been the poet she is. Her aesthetics is the aesthetics of crazy wisdom, the joy that embraces the risk of poetic performance, the aesthetics of all great poets and writers. And unless we as readers can understand that aesthetic, the aesthetic of crazy wisdom, we will not understand, we will ignore, we will underestimate poetry's power to upset conventional wisdom. We need to see and understand the relations and the tensions between them. We need, in other words, both science and art.

Appendix

CONTAINMENT and TRANSFORMATION
in Dickinson's poetry

The following lists provide just a few examples of the schemata operating in characteristic themes in Dickinson's poetry.

PRISON

Confinement – negative

I tried to think a lonelier thing (H85/384, F570/J532/M260)
Lest they should come (A271, F1204/J1169/M557)
The soul has bandaged moments (A85-11/12, F360/J512/M190)
'Twas like a maelstrom (H172, F425/J414/M169)

Confinement – positive

A prison gets to be a friend (H103, F456/J652/M229)
How soft this prison is (H ST22b, F1352/J1334/M585)
Of God we ask one favor (A819, F1675/J1601/M655)
Unto like story (H111, F300/J295/M141)

Escape

Could I but ride indefinite (A88-5/6, F1056/J661/M474)
From all the jails (A189, F1553/J1532/M632)
I never hear the word escape (H272, F144/J77/M81)
Let us play yesterday (H102, F754/J728/M378)
No rack can torture me (H68, F649/J384/M327)
The soul has bandaged moments (A85-11/12, F360/J512/M190)
They shut me up in prose (H182, F445/J613/M223)
What if I say I shall not wait (H69, F305/J277/M159)

SOUL

Containment

The soul selects her own society (H65, F409/J303/M218)
The soul has bandaged moments (A85-11/12, F360/J512/M190)
The soul should always stand ajar (A90-9/10, F1017/J1055/M463)
The soul that hath a guest (A84-1/2, F592/J674/M270)
Exhiliration is within (H67, F645/J383/M325)
Did our best moments last (H116, F560/J393/M285)
A solemn thing within the soul (H106, F467/J483/M234)
The body grows without (H143, F438/J578/M176)
A thought went up my mind (H122, F731/J701/M366)
One blessing had I (H132, F767/J756/M356)
It was a grave (H161, F852/J876/M391, 439)
The props assist the house (H122/347, F729/J1142/M365)
Its hour with itself (H281, F1211/J1225/M502)
I cannot see my soul (A Tr50, F1276/J1262/M709)
There is a solitude of space (H ST5, F1696/J1695/M665)
If ever the lid gets off my head (A Tr18, F585/J1727/M267)
Elysium is as far (A764, F1590/J1760/M726)

Escape

No rack can torture me (H68, F649/J384/M327)
Faith is the peerless bridge (H178, F978/J915/M451)
Twas crisis (A92-1/2, F1093/J948/M488)
Of consciousness her awful mate (A92-15/16, F1076/J894/M482)
Contained in this short life (A159/159a, F1175/J1165/M549)
With pinions of disdain (A529, F1448/J1431/M606)
The overtakelessness of those (H ST22d, F894/J1691/M533)

Transcendence

Dare you see a soul (A162, F401/J365/M214)
Exultation is the going (H254/10, F143/J76/M81)
Safe despair it is that raves (A94-14/15, F1196B/J1243/M555)
There is no frigate (A463, F1286/J1263/M569)
The inundation of the spring (H338, F1423/J1425/M599)

References

Abram, David. 1996. *The Spell of the Sensuous*. New York: Random House.

Aldrich, Thomas Bailey. 1892. "In re Emily Dickinson." *Atlantic Monthly* 69: 143–44. An excerpt entitled "A Poet with no Grammar" is reprinted in Rupp (1972: 18–19).

Anderson, Charles R. 1960. *Emily Dickinson's Poetry: Stairway of Surprise*. New York: Holt, Rinehart and Winston.

Attridge, Derek. 2012. "The Case for the English Dolnik." *Poetics Today* 33 (1): 1–26.

Barlow, Dan, et al. 2014. "The Year in Conferences---2013." *ESQ: A Journal of the American Renaissance* 60: 151–2.

Bates, Arlo. 1970[1890]. "Books and Authors. *Boston Sunday Courier* 96: 2." In *Emily Dickinson: An Annotated Bibliography: Writings, Scholarship, Criticism, and Ana, 1850–1968*, edited by Willis J. Buckingham, 28–33. Bloomington and Indiana: Indiana University Press.

Bauer, Matthias, Sigrid Beck, Saskia Brockmann, Susanne Riecker, Angelika Zirker, and Nadine Bade. 2020. *Linguistics Meets Literature: More on the Grammar of Emily Dickinson*. Berlin: De Gruyter Mouton.

Berlin, Adele. 1985. *The Dynamics of Biblical Parallelism*. Bloomington: Indiana University Press.

Bianchi, Martha D., and Alfred L. Hampson, eds. 1935. *Unpublished Poems of Emily Dickinson*. Boston: Little, Brown.

Blake, William. 1972[1810]. *Blake: Complete Writings with Variant Readings*, edited by Geoffrey Keynes. London and New York: Oxford University Press.

Brandt, Per Aage. 2004. *Spaces, Domains, and Meaning: Essays in Cognitive Semiotics*. European Semiotics Series No. 4. Bern: Peter Lang.

Brooks, Cleanth, Richard W. B. Lewis, and Robert Penn Warren, eds. 1973. *American Literature: The Makers and the Making*. New York: St. Martin's Press.

Brown, Penelope, and Stephen Levinson. 1978. "Universals in Language Usage: Politeness Phenomena." In *Questions and Politeness: Strategies in Social Interaction*, edited by Esther N. Goody, 56–324. Cambridge: Cambridge University Press.

Buckingham, Willis J., ed. 1970. *Emily Dickinson: An Annotated Bibliography: Writings, Scholarship, Criticism, and Ana, 1850–1968*. Bloomington and Indiana: Indiana University Press.

Budick, Emily M. 1985. *Emily Dickinson and the Life of Language*. Baton Rouge and London: Louisiana State University Press.

Burman, Edward. 1984. *The Inquisition: Hammer of Heresy*. New York: Dorset Press.

Bybee, Joan L. 1985. *Morphology: A Study of the Relation Between Meaning and Form*. Amsterdam: John Benjamins.

Cameron, Sharon. 1992. *Choosing Not Choosing: Dickinson's Fascicles*. Chicago and London: The University of Chicago Press.

Cameron, Sharon. 1979. *Lyric Time: Dickinson and the Limits of Genre*. Baltimore and London: The Johns Hopkins University Press.

Carlisle, Thomas. 1987. *Invisible Harvest*. Grand Rapids: William B. Eerdmans.

Carse, James P. 2008. *The Religious Case Against Belief*. New York and London: Penguin Books.

Cody, John. 1970. *After Great Pain: The Inner Life of Emily Dickinson*. Cambridge, MA: The Belknap Press of Harvard University Press.

Creed, Robert P. 1990. *Reconstructing the Rhythm of Beowulf*. Columbia: University of Missouri Press.

Crumbley, Paul. 1997. *Inflections of the Pen: Dash and Voice in Emily Dickinson*. Lexington: The University Press of Kentucky.

Cureton, Richard. 1992. *Rhythmic Phrasing in English Verse*. London: Longman.

Damasio, Antonio. 1999. *The Feeling of What Happens: Body and Emotion in the Making of Consciousness*. New York: Harcourt Brace.

Damasio, Antonio. 1994. *Descartes' Error: Emotion, Reason, and the Human Brain*. New York: Putnam.

Darwin, Charles. 1859. *On the Origin of Species*. London: John Murray.

Dill, Lesley. 2018. *Divide Light*. https://www.lesleydill.net/divide-light-opera.

Duchac, Joseph. 1979. *The Poems of Emily Dickinson: An Annotated Guide to Commentary Published in English, 1890–1977*. New York: G. K. Hall.

Dyson, Freeman. 1980. *Infinite in All Directions*. New York: Harper and Row.

Eberwein, Jane Donahue. 1998. *An Emily Dickinson Encyclopedia*. Westport: Greenwood Press.

Eberwein, Jane Donahue. 1985. *Dickinson: Strategies of Limitation*. Amherst: The University of Massachusetts Press.

Esparza, Daniel. 2018. "The 'Devil's Chord': A Forbidden Medieval Musical Sequence." Accessed June 30, 2022. https://aleteia.org/2018/10/25/a-medieval-forbidden-musical -sequence-the-devil-in-music-or-the-devils-chord/.

Farr, Judith. 1992. *The Passion of Emily Dickinson*. Cambridge, MA and London: Harvard University Press.

Fauconnier, Gilles. 1997. *Mappings in Thought and Language*. Cambridge: Cambridge University Press.

Fauconnier, Gilles. 1985. *Mental Spaces*. Cambridge, MA: The MIT Press. Revised edition. New York: Cambridge University Press, 1994.

Fauconnier, Gilles, and Mark Turner. 2002. *The Way We Think: Conceptual Blending and the Mind's Hidden Complexities*. New York: Basic Books.

Fauconnier, Gilles, and Mark Turner. 1994. *Conceptual Projection and Middle Spaces*. Cognitive Science Report 9401, University of California/San Diego.

Ferris, Timothy. 1988. *Coming of Age in the Milky Way*. New York: William Morrow and Company.

Fillmore, Charles J. 1977. "Topics in Lexical Semantics." In *Current Issues in Linguistic Theory*, edited by R. W. Cole, 76–138. Bloomington: Indiana University Press.

Finnerty, Paraic. 2014. "'It Does Not Mean Me, but a Supposed Person': Browning, Dickinson, and the Dramatic Lyric." *Comparative American Studies* 12 (4): 264–281.

Fludernik, Monika. 2019. *Metaphors of Confinement: The Prison in Fact, Fiction, and Fantasy*. Oxford: Oxford University Press.

Foolen, Ed. 2004. "Expressive Binomial NPs in Germanic and Romance Languages." In *Studies in Linguistic Motivation*, edited by Günter Radden and Klaus-Uwe Panther. Cognitive Linguistics Research 28. Berlin and New York: Mouton de Gruyter.

Forgas, Joseph P., ed. 2000. *Feeling and Thinking: The Role of Affect in Social Cognition*. Cambridge: Cambridge University Press.

Franklin, Ralph W., ed. 1998. *The Poems of Emily Dickinson: Variorum Edition*. 3 vols. Cambridge, MA: The Belknap Press of Harvard University Press.

Franklin, Ralph W., ed. 1981. *The Manuscript Books of Emily Dickinson*. Cambridge, MA: The Belknap Press of Harvard University Press.

Freeman, Donald C. 1978. "Keats's 'To Autumn': Poetry as Process and Pattern." *Language and Style* 111: 3–17.

Freeman, Margaret H. 2020a. *The Poem as Icon: A Study in Aesthetic Cognition*. New York: Oxford University Press.

Freeman, Margaret H. 2020b. "On the Measures of English Verse." *Herald of Kyiv National Linguistic University. Series in Philology* 23 (1): 8–19.

Freeman, Margaret H. 2008. "Reading Readers Reading a Poem: From Conceptual to Cognitive Integration." *Cognitive Semiotics* 2: 102–128.

Freeman, Margaret H. 2006. "From Metaphor to Iconicity in a Poetic Text." In *The Metaphors of Sixty: Papers Presented on the Occasion of the 60th Birthday of Zoltán Kövecses*, edited by Réka Benczes and Szilvia Csábi, 127–135. Budapest: Eötvös Loránd University.

Freeman, Margaret H. 2005a. "Is iconicity Literal? Cognitive Poetics and the Literal Concept in Poetry." In *The Literal and Nonliteral in Language and Thought*, edited by Seana Coulson and Barbara Lewandowska-Tomaszczyk, 65–83. Frankfurt am Main: Peter Lang.

Freeman, Margaret H. 2005b. "Poetry as Power: The Dynamics of Cognitive Poetics as a Scientific and Literary Paradigm." In *Cognition and Literary Interpretation in Practice*, edited by Harri Veivo, Bo Pettersson, and Merja Polvinen, 31–57. Helsinki: Helsinki University Press.

Freeman, Margaret H. 2002a. "Cognitive Mapping in Literary Analysis." *Style* 36 (3): 466–83.

Freeman, Margaret H. 2002b. "Momentary Stays, Exploding Forces: A Cognitive Linguistic Approach to the Poetics of Emily Dickinson and Robert Frost." *Journal of English Linguistics* 30 (1): 73–90.

Freeman, Margaret H. 2002c. "The Body in the Word: A Cognitive Approach to the Shape of a Poetic Text." In *Cognitive Stylistics: Language and Cognition in Text Analysis*, edited by Elena Semino and Jonathan Culpeper, 23–47. Amsterdam/Philadelphia: John Benjamins Publishing Company.

Freeman, Margaret H. 2000. "Poetry and the Scope of Metaphor: Toward a Cognitive Theory of Literature." In *Metaphor and Metonymy at the Crossroads*, edited by Antonio Barcelona, 253–81. Berlin: Mouton.

Freeman, Margaret H. 1996. "Emily Dickinson and the Discourse of Intimacy." In *Semantics of Silences in Linguistics and Literature*, edited by Gudrun M. Grabher and Ulrike Jeßner, 191–210. Heidelberg: Universitätsverlag C. Winter.

Freeman, Margaret H. 1995. "Metaphor Making Meaning: Emily Dickinson's Conceptual Universe." *Journal of Pragmatics* 24: 643–66.

Freeman, Margaret H., and Nigel McLoughlin. 2021. "'To Pile Like Thunder': The Advantages of Reading Emily Dickinson's Poetry from a Cognitive Perspective." *The Emily Dickinson Journal* 30 (1): 1–27.

Freeman, Margaret H., and Masako Takeda. 2006. "Art, Science, and Ste. Emilie's Sunsets: A Háj-inspired Cognitive Approach to Translating an Emily Dickinson Poem into Japanese. Festschrift for John Robert Ross." *Style* 40 (1–2): 109–127.

Frost, Robert. 1972. "The Figure A Poem Makes." In *Robert Frost: Poetry and Prose*, edited by Edward Connery Latham and Lawrance Thompson, 393–396. New York: Holt, Rinehart and Winston.

Fry, Hannah. 2021. "What Really Counts." *The New Yorker*, March 29: 70–73.

Gardner, R. Allen, Beatrix T. Gardner, and Thomas E. Van Cantfort. 1989. *Teaching Sign Language to Chimpanzees*. Albany: State University of New York Press.

Gasparov, Mikhail L. 1996. *A History of European Versification*, edited by G. S. Smith and L. Holford-Strevens, translated by G. S. Smith and M. Tarlinskaja. Oxford: The Clarendon Press; New York: Harper and Row.

Gibbs, Raymond W., Jr. 1999. *Intentions in the Experience of Meaning*. Cambridge: Cambridge University Press.

Gibbs, Robert. 1991. "The Other Comes to Teach Me: A Review of Recent Levinas Publications." *Man and World: An International Philosophical Review* 24: 219–233.

Gibson, James J. 1966. *The Senses Considered as Perceptual Systems*. London: Allen and Unwin.

Grabher, Gudrun M. 2019. *Levinas and the Other in Narratives of Facial Disfigurement: Singing Through the Mask*. London and New York: Routledge.

Hagenbüchle, Roland. 1986. "Sign and Process: The Concept of Language in Emerson and Dickinson." *Dickinson Studies* 58: 59–88.

Haiman, John, ed., 1985. *Iconicity in Syntax*. Typological Studies in Language (TSL). Amsterdam/Philadelphia: John Benjamins Publishing Company.

Haley, Michael Cabot. 1988. *The Semeiosis of Poetic Metaphor*. Bloomington and Indiana: Indiana University Press.

Halle, Morris, and Samuel Jay Keyser. 1971. *English Stress: Its Form, its Growth, and its Role in Verse*. New York: Harper and Row.

Halliday, M. A. K. 1964. *Descriptive Linguistics in Literary Studies*. Edinburgh: Edinburgh University Press.

Hamill, Sam, ed. 2003. *Poets Against the War*. New York: Thunder's Mouth Press/Nation Books.

Hart, Ellen Louise, and Martha Nell Smith. 1998. *Open Me Carefully: Emily Dickinson's Intimate Letters to Susan Huntington Dickinson*. Ashfield: Paris Press.

Hascall, Dudley L. 1974. "Triple Meter in English Verse." *Poetics* 3 (4): 49–71.

Hawthorne, Nathaniel. 1922[1851]. *The House of the Seven Gables*, edited by A. Marion Merrill. Boston: Allyn and Bacon.

Heginbotham, Eleanor E. 2003. *Reading the Fascicles of Emily Dickinson: Dwelling in Possibilities*. Columbus: The Ohio State University Press.

Hinton, David. 2019. *Awakened Cosmos: The Mind of Classical Chinese Poetry*. Boulder: Shambhala.

Hiraga, Masako K. 2005. *Metaphor and Iconicity: A Cognitive Approach to Analysing Texts*. Houndsmill, Basingstoke, and New York: Palgrave Macmillan.

Hiraga, Masako K. 1998. "Metaphor-Icon Link in Poetic Texts: A Cognitive Approach to Iconicity." *Journal of the University of the Air* 16: 95–123.

Hogue, Cynthia. 1961. "Dickinson's 'I heard a fly buzz when I died.'" *The Explicator* 20 (3): item 26.

Holland, Norman. 1988. *The Brain of Robert Frost: A Cognitive Approach to Literature*. New York and London: Routledge, Chapman, and Hall.

Holyoak, Keith J., and Paul Thagard. 1995. *Mental Leaps: Analogy in Creative Thought*. Cambridge, MA and London: The MIT Press.

Howe, Susan. 1987. "Some Notes on Visual Intentionality in Emily Dickinson." Accessed June 30, 2022. https://www.asu.edu/pipercwcenter/how2journal/archive/print_archive/alertsvol3no4.html#some.

Howe, Susan. 1985. *My Emily Dickinson*. Berkeley: North Atlantic Books.

Ingarden, Roman. 1973. *The Cognition of the Literary Work of Art*, translated by Ruth Ann Crowley and Kenneth R. Olson. Evanston: Northwestern University Press.

Ivanov, Vyacheslav Vsevolodovich. 2004. "O printsipakh russkogo stikha [Principles of Russian Versification]." In *Analysieren als Deuten: Wolf Schmid zum 60*, edited by Lazar Fleishman, Christine Gölz, and Aage A. Hansen-Löve, 97–110. Geburtstag. Hamburg: Hamburg University Press.

Ivanov, Vyacheslav Vsevolodovich. 1996. "Unstressed Intervals in Brodsky's Dol'niki." *Elementa* 2 (3/4): 277–284.

Jevons, William Stanley. 1870. *Elementary Lessons in Logic: Deductive and Inductive*. Oxford and London: Macmillan.

Johnson, Greg. 1985. *Emily Dickinson: Perception and the Poet's Quest*. Alabama: University of Alabama Press.

Johnson, Mark. 1987. *The Body in the Mind: The Bodily Basis of Meaning, Imagination, and Reason*. Chicago: The University of Chicago Press.

Johnson, T. H., ed. 1965. *The Letters of Emily Dickinson*. Cambridge, MA: The Belknap Press of Harvard University Press.

Johnson, T. H., ed. 1955. *The Poems of Emily Dickinson: Including Variant Readings Critically Compared with All Known Manuscripts*. Cambridge, MA: The Belknap Press of Harvard University Press.

Juhasz, Suzanne. 1983. *The Undiscovered Continent: Emily Dickinson and the Space of the Mind*. Bloomington: Indiana University Press.

Kang, Yanbin. 2021. "Dickinson's Daisy/Sun (set), Daoism, and Emerson." *The Explicator* 79 (1–2): 41–47.

Kant, Immanuel. 1934[1787]. *Critique of Pure Reason*. 2nd edition, edited by J. M. D. Meiklejohn. London: J. M Dent; New York: E. P. Dutton.

Keats, John. 1819. "Ode to a Nightingale." In *The Oxford Book of English Verse: 1250–1900*, edited by Arthur Quiller-Couch. Oxford: Clarendon, 1919.

Kenny, Virginia. 1960. "Was Emily Dickinson Short of Hearing?" In *Highlights, Bulletin of the New York League for the Hard of Hearing* XXXIX: 5, 11, 15.

Kermode, Frank. 1967. *The Sense of an Ending: Studies in the Theory of Fiction*. Oxford: Oxford University Press.

Kher, Inder N. 1974. *The Landscape of Absence: Emily Dickinson's Poetry*. New Haven: Yale University Press.

Kövesces, Zoltán. 2000. *Metaphor and Emotion: Language, Culture, and Body in Human Feeling*. Cambridge: Cambridge University Press.

Kövesces, Zoltán. 1986. *Metaphors of Anger, Pride, and Love: A Lexical Approach to the Structure of Concepts*. Amsterdam: John Benjamins.

Kowalchik, C., and W. H. Hylton, eds. 1987. *Rodale's Illustrated Encyclopedia of Herbs*. Emmaus: Rodale Press.

Lakoff, George, and Mark Johnson. 1998. *Philosophy in the Flesh*. Chicago and London: University of Chicago Press.

Lakoff, George, and Mark Johnson. 1980. *Metaphors We Live By*. Chicago and London: The University of Chicago Press.

Langacker, Ronald W. 1987, 1991. *Foundations of Cognitive Grammar*. 2 vols. Stanford: Stanford University Press.

Langer, Susanne K. 1967. *Mind: An Essay on Human Feeling*. Baltimore: Johns Hopkins University Press.

Langer, Susanne K. 1953. *Feeling and Form: A Theory of Art*. New York: Charles Scribner's.

Leão, Lucia. 2002. "The Labyrinth as a Model of Complexity: The Semiotics of Hypermedia." *Computational Semiotics for New Media* (COSIGN). https://www.academia.edu/952859/The_labyrinth_as_a_model_of_complexity_the_semiotics_of_hypermedia.

Lebow, Lori. 1999. *Autobiographic Self-Construction in the Letters of Emily Dickinson*. Unpublished Ph.D. dissertation, University of Wollongong.

Leder, Drew. 1990. *The Absent Body*. Chicago: The University of Chicago Press.

LeDoux, Joseph. 1996. *The Emotional Brain: The Mysterious Underpinnings of Emotional Life*. New York: Simon and Schuster.

Leiter, Sharon. 2007. *Critical Companion to Emily Dickinson: A Literary Reference to Her Life and Work*. New York: Facts on File.

Levin, Samuel R. 1971. "The Analysis of Compression in Poetry." *Foundations of Language* 7 (1): 38–55.

Lewes, George Henry. 1879. *Problems of Life and Mind: The Study of Psychology—Its Object, Scope, and Method*. London: Trübner & Co., Ludgate Hill.

Leyda, Jay. 1960. *The Years and Hours of Emily Dickinson*. 2 vols. New Haven: Yale University Press.

Loeffelholz, Mary. 2005. "Dickinson's Decoration." *ELH* 72 (3): 663–89. Accessed June 30, 2022. http://www.jstor.org/stable/30030068.

Lucas, Dolores D. 1969. *Emily Dickinson and Riddle*. Dekalb: Northern Illinois University.

MacGregor, Jenkins. 1930. *Emily Dickinson: Friend and Neighbor*. Boston: Little, Brown, and Company.

MacLeish, Archibald. 1960. *Poetry and Experience*. Boston: Houghton Mifflin.

McCarthy, Suzanne. 2002. "Giselle." In *Insight Day* (March issue). London: Linbury Studio Theatre, Royal Opera House.

McCawley, James D. 1981. *Everything that Linguists Have Always Wanted to Know about Logic (But Were Ashamed to Ask)*. Chicago: The University of Chicago Press, and Oxford: Basil Blackwell.

McLoughlin, Nigel F. 2017. Shape-Shifting Instabilities: Using Blending Theory to Understand Metamorphosis in Eavan Boland's Poetry." In *Eavan Boland: Inside History*, edited by Siobhan Campbell and Nessa O'Mahony, 195–218. Dublin: Arlen House.

McLoughlin, Nigel F. 2016. "Into the Futures of their Makers: A Cognitive Poetic Analysis of Reversals, Accelerations and Shifts in Time in the Poems of Eavan Boland." In *World Building: Discourse in the Mind. Advances in Stylistics*, edited by Joanna Gavins and Ernestine Lahey. London: Bloomsbury Academic.

Merleau-Ponty, Maurice. 1968. *The Visible and the Invisible*, edited by Claude Lefort; translated by Alphonso Lingis. Evanston: Northwestern University Press.

Merleau-Ponty, Maurice. 1962. *Phenomenology of Perception*, translated by Colin Smith. London: Routledge.

Miall, David S. 2006. *Literary Reading: Empirical and Theoretical Studies*. New York: Peter Lang.

Miller, Cristanne, ed. 2016. *Emily Dickinson's Poems: As She Preserved Them*. Cambridge, MA and London: The Belknap Press of Harvard University Press.

Miller, Cristanne. 2012. *Reading in Time: Emily Dickinson in the Nineteenth Century*. Amherst and Boston: University of Massachusetts Press.

Miller, Ruth. 1968. *The Poetry of Emily Dickinson*. Middletown: Wesleyan University Press.

Mitchell, Domhnall. 2005. *Measures of Possibility: Emily Dickinson's Manuscripts*. Amherst and Boston: University of Massachusetts Press.

Moore, Geoffrey. 1964. *American Literature: A Representative Anthology from Colonial Times to the Present*. London: Faber and Faber.

Mukarovský, Jan. 1970. "Standard Language and Poetic Language," edited and translated by P. L. Garvin." In *Linguistics and Literary Style*, edited by Donald C. Freeman, 40–56. New York: Holt, Rinehart and Winston.

Neisser, Ulric. 1976. *Cognition and Reality*. San Francisco: Freeman.

Newen, Albert, Leon De Bruin, and Shaun Gallagher, eds. 2018. *The Oxford Handbook of 4E Cognition*. London and New York: Oxford University Press.

Noë, Alva. 2009. *Out of Our Heads: Why You Are Not Your Brain, and Other Lessons from the Biology of Consciousness*. New York: Farrar, Straus and Giroux.

Nordquist, Richard. 2021. "Sound Symbolism in English: Definition and Examples." ThoughtCo, March 24. Accessed June 30, 2022. thoughtco.com/sound-symbolism-words-1692114.

Oberhaus, Dorothy H. 1995. *Emily Dickinson's Fascicles: Method and Meaning*. University Park: The Pennsylvania State University Press.

O'Keefe, Martha. 1986. *This Edifice: Studies in the Structure of the Fascicles of the Poetry of Emily Dickinson*. Privately printed.

Olney, James. 1993. *The Language of Poetry: Walt Whitman, Emily Dickinson, Gerard Manley Hopkins*. Athens: University of Georgia Press.

Owen, Stephen. 1985. *Traditional Chinese Poetry and Poetics*. Madison: The University of Wisconsin Press

Pagán Cánovas, Cristóbal. 2011. "The Genesis of the Arrows of Love: Diachronic Conceptual." in Greek Mythology. *American Journal of Philology* 132: (4): 553–579.

Parker, Alice. 2020. *The Gift of Song*. Chicago: GIA Publications.

Patterson, Rebecca. 1979. *Emily Dickinson's Imagery*, edited by Margaret H. Freeman. Amherst: University of Massachusetts Press.

Patterson, Rebecca. 1951. *The Riddle of Emily Dickinson*. Boston: Houghton Mifflin.

Patterson, Richard Ferrar, ed. 1974. *Ben Jonson's Conversations with William Drummond of Hawthornden*. New York: Haskell House Publishers.

Peirce, Charles S. 1935, 1938. *Collected Papers of Charles Sanders Peirce*. Cambridge, MA: Harvard University Press.

Pope, Alexander. 1924[1711]. "An Essay on Criticism." In *Alexander Pope's Collected Poems*, edited by Bonamy Dobrée, 58–76. London: J. M. Dent and Sons.

Porter, David. 1981. *Dickinson: The Modern Idiom*. Cambridge, MA: Harvard University Press.

Porter, David. 1966. *The Art of Emily Dickinson's Early Poetry*. Cambridge, MA: Harvard University Press.

Porter, Ebenezer. 1835. *The Rhetorical Reader: Instructions for Regulating the Voice with a Rhetorical Notation*. New York: Mark H. Newman.

Premack, David, and Guy Woodruff. 1978. "Does the Chimpanzee Have a Theory of Mind?" *Behavioral and Brain Sciences* 4 (4): 515–629.

Rich, Adrienne. 1970. "Vesuvius at Home: The Power of Emily Dickinson." *Parnassus: Poetry in Review* 15 (Fall/Winter 1976): 49–74.

Robbins, Tom. 2004. "In Defiance of Gravity: Writing, Wisdom, and the Cabulous club Gemini." *Harper's Magazine*, September: 57–61.

Robertson, Nan. 1980. *The New York Times*, November 5, 1980: C25.

Rosenbaum, Samuel P., ed. 1964. *A Concordance to the Poems of Emily Dickinson*. Ithaca: Cornell University Press.

Ross, John Robert (Haj). 2008. "Structural Prosody." *Cognitive Semiotics* 2: 65–82.

Ross, John Robert (Haj). 2000. "The Taoing of a Sound – Phonetic Drama in William Blake's 'The Tyger.'" In *Phonosymbolism and Poetic Language*, edited by Patrizia Violi, 99–145. Turnhout, Belgium: Brepols.

Rupp, Richard H, ed. 1972. *Critics on Emily Dickinson*. Coral Gables: University of Miami Press.

Salska, Agnieska. 1985. *Walt Whitman and Emily Dickinson: Poetry of the Central Consciousness*. Philadelphia: University of Pennsylvania Press.

Selkirk, Elizabeth. 1984. *Phonology and Syntax: The Relation Between Sound and Structure*. Cambridge, MA: The MIT Press.

Sewall, Richard B. 1974. *The Life of Emily Dickinson*, 2 vols. New York: Farrar, Straus and Giroux.

Sewall, Richard B. 1948. "Dickinson's 'To undertake is to achieve.'" *Explicator* VI, Item 51. Reprinted in 1966. "Dickinson," *The Explicator Cyclopedia*, Vol 1, *Modern Poetry*, edited by Walcutt, Charles Child, and Edwin Whitesell, 55–88.

Shakespeare, William. 1954. *The Complete Works of William Shakespeare*, edited by W. J. Craig. London: Oxford University Press.

Shapiro, Susanne. 2001. "Secrets of the Pen: Emily Dickinson's Handwriting." In *Emily Dickinson at Home*, edited by Gudrun M. Grabher and Martina Antretter, 223–238. Trier: Wissenschaftlicher Verlag Trier.

Shelley, Percy Bysshe. 1900. *Poems of Shelley: Selected and Arranged by Stopford A. Brooke*. London: MacMillan.

Small, Judy Jo. 1990. *Positive as Sound: Emily Dickinson's Rhyme*. Athens: University of Georgia Press.

Stevens, Wallace. 1957 "The Irrational Element in Poetry." *Opus Posthumous*, 216–229. New York: Alfred A. Knopf.

Stonum, Gary. 1990. *The Dickinson Sublime*. Madison: The University of Wisconsin Press.

Ströbel, Liane. 2014. "There is Something about Fear...." In *Meaning, Frames, and Conceptual Representation*, edited by T. Gamerschlag, D. Gerland, R. Osswald, and W. Petersen, 143–160. Düsseldorf: dup.

Talmy, Leonard. 2000. *Toward a Cognitive Semantics*. 2 vols. Cambridge, MA: The MIT Press.

Tarlinskaja, Marina. 1992. "Metrical Typology: English, German, and Russian Dolnik Verse." *Comparative Literature* 44 (1): 1–21.

Todd, Mabel Loomis, and Millicent Todd Bingham, eds. 1945. *Bolts of Melody: New Poems of Emily Dickinson*. New York: Harper and Brothers.

Tsur, Reuven, and Chen Gafni. In press. *Sound–Emotion Interaction in Poetry: Rhythm, Phonemes, Voice Quality*. Amsterdam/Philadelphia: John Benjamins Publishing Company.

Ungerer, F., and H. J. Schmid. 1996. *An Introduction to Cognitive Linguistics*. London: Longman.

Vendler, Helen. 2010. *Dickinson: Selected Poems and Commentaries*. Cambridge, MA: The Belknap Press of Harvard University Press.

Venn, John. 1880. "On the Diagrammatic and Mechanical Representation of Propositions and Reasonings." *London, Edinburgh & Dublin Philosophical Magazine and Journal of Science* 10 (59): 1–18.

Vico, Giambattista. 1948[1744]. *The New Science of Giambattista Vico* (Principi di scienza nuova d'intorno alla communinatura delle nazioni), translated by Max Harold Fisch and Thomas Goddard Bergin. Ithaca: Cornell University Press.

Virgil. 1910. *Aeneid*, translated by Theodore C. Williams. Boston: Houghton Mifflin.

Waggoner, Hyatt. 1968. *American Poets: From the Puritans to the Present*. Boston: Houghton Mifflin.

Wardrop, Daneen. 1996. *Emily Dickinson's Gothic: Goblin with a Gauge*. Iowa City: University of Iowa Press.

Webster, Noah. 1844. *An American Dictionary of the English Language*. 2 vols. Amherst, MA: J. S. and C. Adams.

Weisbuch, Robert. 1975. *Emily Dickinson's Poetry*. Chicago and London: The University of Chicago Press.

Werner, Marta L., ed. 1995. *Emily Dickinson's Open Folios: Scenes of Reading, Surfaces of Writing*. Ann Arbor: University of Michigan Press.

Wordsworth, William. 1904[1798]. *Poetical Works*, edited by Thomas Hutchinson and Ernest De Selincourt. London: Oxford University Press.

Worrall, Simon. 2002. *The Poet and the Murderer: The True Story of Literary Crime and the Art of Forgery*. New York: Dutton.

Wylder, Edith. 1971. *The Last Face: Emily Dickinson's Manuscripts*. Albuquerque: University of New Mexico Press.

Zlatev, Jordan. 1997. *Situated Embodiment: Studies in the Emergence of Spatial Meaning*. Stockholm: Gotab.

Index of First Lines

General Index